The Keeney Place

A Life in the Heartland

Dennis R. Keeney

The Keeney Place

K^{The}eeney Place

A Life in the Heartland

Dennis R. Keeney

Foreword by
Paul W. Johnson

Cover photograph by Jim Heemstra and used with permission
    from Iowa State Alumni Association.
Illustrations by Anika Carlson.
Author photograph by Jacquelyn's Photography.
Text design by Connie Kuhnz.
Composition by Bookmobile Design and Digital Publisher Services,
    Minneapolis, MN.

Levins Publishing
For individual online orders go to: www.TheKeeneyPlace.com
Bulk orders and bookstores contact Levins Publishing at:
    info@LevinsPublishing.com

ISBN 978-0-9853972-5-8
Library of Congress Control Number: 2014959285
Printed in the United States

For my parents, who gave me my love of farming,
for my wife Betty,
who stood by me throughout my journey home,
and for my children and grandchildren,
who give me hope.

This land is your land. This land is my land.

WOODY GUTHRIE

ACKNOWLEDGMENTS

So many wonderful people have helped and influenced me over the years that I won't even try to name them here. I have mentioned many of them in the text of this book and hope the others understand that I appreciate them more than they will ever know.

I do, however, want to acknowledge the 1986 and 1987 Iowa General Assembly that worked so tirelessly to pass the Iowa Groundwater Protection Act and establish the Leopold Center for Sustainable Agriculture. Their vision and courage provided me the opportunity of a lifetime to help a new generation of farmers find their way.

Finally, I acknowledge the editing and design expertise of Jane Dickerson and Alexandra Erickson of Levins Publishing. Without their support and encouragement, this book would never have been completed.

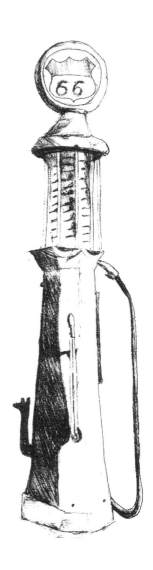

In 1862 President Abraham Lincoln signed into law the Morrill Act. It offered states 30,000 acres of land for each of their Senators and Representatives. The land was to be sold and its proceeds used to establish colleges in each state to provide higher education for the "industrial classes." These institutions became known as "land-grant colleges," and today every state in the Union has at least one land-grant university. In 1887 the Hatch Act added research, and in 1914 the Smith-Lever Act added an extension component. Today, land-grant universities, with their education, extension, and research components can be credited with one of the most revolutionary changes in the status of humanity that our world has ever witnessed.

What does this have to do with *The Keeney Place: A Life in the Heartland*? Everything. Dennis Keeney starts his book by writing about his childhood on an Iowa farm, a mid-twentieth century product of the land-grant idea. And what a childhood. Any of us who grew up in rural America at that time can't help but reminisce as we read his words. Dennis writes about his Grandpa on his new-fangled tractor hollering "whoa" as the tractor raced out of control down the slope into the farm pond. Change came lickety-split, often faster than many of our grandparents could handle. Overall it was a good era. I sometimes mourn the loss of those experiences, which formed a whole generation of Americans.

In Keeney's second section he becomes the raw material for which the land-grants were created. He, and his future partner, Betty, enrolled at Iowa State College, the first land-grant, established in 1862. Dennis describes for us a life-changing experience, an introduction to the noble pursuit of advancing the science of food production, and land stewardship. His education was liberal and liberating. What an exciting time. The human condition was improving, and agriculture was at the center of it.

Dennis continued in the land-grant tradition and tells of his notable career in agricultural research, education, and extension at the University of Wisconsin. But it was also a time when many, both in and out of the academy, began to question the direction of our drive for progress. Early on in this era, Dennis Keeney became aware of this uncertainty. Were the land-grants unable to accept criticism? Had they lost their way? Were they unduly influenced by profit over principle? Rachel Carson made it clear in her book *Silent Spring* that not all agricultural science was noble science. The land-grants fought her tooth and nail. Dennis tells us of his efforts to question where agricultural science was taking us.

After his land-grant education and research career, Dennis tells of his return to Iowa to lead the newly established Leopold Center for Sustainable Agriculture at Iowa State University. Needless to say, when I learned of the Keeney appointment I was pleased.

I was a farmer-legislator in Iowa at that time and helped develop the Leopold Center as a part of the 1987 Groundwater Protection Act. We had serious concerns about locating it at Iowa State. Iowa was still in the middle of the farm crisis of the 1980s, and land-grant universities were rapidly losing their luster. In the end, we decided to place it at Iowa State with the hope that the Leopold Center would give faculty and students an alternative to the industrial agriculture approach. Chuck Benbrook, the Executive Director of the National Academy of Sciences Board on Agriculture at the time, called the Leopold Center "the land-grant university's last hope."

No person was better equipped to develop the Leopold Center than Dennis Keeney. He was not an outsider. He was a respected researcher. But more important, he was a product of the land-grant idea and was deeply concerned about its loss of tradition and its apparent inability to develop a renewed vision of service.

All is not well today, but partly as a result of Dennis Keeney's efforts there is hope. Sustainable agriculture programs are developing throughout the country, many by non-governmental institutions. At many land-grant universities, faculty and students are questioning the thinking that bigger industrial agriculture is better agriculture. However, many university administrators have yielded to pressure from their industrial partners and thus far have resisted the Leopold Center concept.

Dennis Keeney's willingness to question blind profit-driven approaches to agricultural biotechnology, biofuels, and large animal confinement systems should be admired in education and research circles. Unfortunately, land-grant universities often feel threatened by those who would question corporate-driven research. The response of agriculture to criticism always seems to begin with denial (antibiotic resistance hasn't been caused by overuse in animal feeds, or hypoxia in the Gulf of Mexico isn't caused by farming practices in the farm belt). Then the response moves to scare tactics about starvation (the green revolution's dependence on overuse of fertilizers, pesticides, and new, patented seed varieties have saved millions from death). Yet, I still believe that when they set their minds to it, our land-grant universities are capable of developing appropriate alternatives.

The Keeney Place: A Life in the Heartland is a wonderful story of how a farm boy from Iowa, formed by a great idea, can give us a better world. Think of how much better it could be if we'd all pause, reflect on this good life, and recommit to the idea envisioned in the 1862 Morrill Act: A world open to liberal education, good objective science, and a commitment to service, and respect for this wonderful world we live in.

PAUL W. JOHNSON

The Keeney Place, as I affectionately call the farm I grew up on, was a victim of the times. Doomed by untimely advice from the land grant university and its industrial partners, the farm became too much for Dad and Mom to bear. They hung on for several years after I left for college, and then it was sold to an abusive land baron. Thus began our home farm's slide from the pride of our family to a small part of an enormous agricultural industrial park we call the Corn Belt. I went on to be a university professor, Dad died, and a tornado destroyed the grand old barn that stood sentinel over the property.

When Dad died in sorrow because he had to leave the home place, I realized just how much I loved the Keeney Place of my memory. My life story over the past fifty years has been an often-stumbling attempt to get back there. In that sense, I'm not that different from most of you who will read this book. No matter where we are now, we all have, somewhere in our past, connections to the land. Even though we can't go back, we continue to care about the land as if it were still in the family.

Aldo Leopold wrote in his *Sand County Almanac* that, "When we see land as a community to which we belong we may begin to use it with love and respect. There is no other way for land to survive the impact of mechanized man, nor for us to reap from it the aesthetic harvest it is capable, under science, of contributing to our culture." I like this passage a lot. It took me many, many years to fully appreciate the idea of farmland as a community that involves each of us. No wonder it's called "The Heartland." But it's a land in deep trouble, ruled by industrialization, and moving more distant from Leopold's vision every day. Only community, our community, can set things right.

I invite you to read the story of how I found my way back to the Keeney Place. I hope it inspires you to find your way back as well.

The Keeney Place

A Life in the Heartland

Runnells High School

CHAPTER I

Growing Up on an Iowa Farm

I was born in 1937 in a small hospital not too far from my grandfather's farm near Woodburn in southern Iowa. My mother kept a small diary of those days. One short sentence from it is all you need to know about that farm: "To make a living was hard." Depression-level crop prices, not nearly enough rain, and plagues of grasshoppers and cinch bugs were "an army mowing everything down," making life on the farm about as far from idyllic as one can imagine. Add to the mix soil more suited to the shrubs that gave the area its "Puckerbrush" name than to rich stands of corn, and you can easily see why Mother eventually accepted the inevitable and started looking for a better place to live. She and Dad settled on a farm they could afford near Runnells, Iowa, a small town about twelve miles east of Des Moines. We moved in 1941 to what then was called the Warren Place. Then, in 1945, Grandpa, Grandma, and Uncle Morris moved to Vandalia, about six miles from our farm, so they could farm together with Dad and share equipment and labor. By the time I started kindergarten, my family had found a place that could, with lots of hard work and, with a little luck, provide Dad, Mom, me, and my sisters Patty and Janis, who were to come along later, with a modest life. This was the farm I grew up on, and the one I remember as though I were still there.

As the years went by, the Webster Place gradually became known as the Keeney Place. That's the way it works in rural areas— you have to live somewhere for a while before the locals accept you enough to associate your name with where you live. The Keeney Place had a house, outbuildings, and a glorious barn that was a landmark in the region. Constructed of hollow tiles, two layers thick, and four stories high, it perched on a ridge and provided the perfect vantage for a young boy to look in all directions and dream of the future. The farm itself was 200 acres in size, a little smaller than the total incorporated area of nearby Runnells. The farm had been blessed with fertile soils that bring to mind such phrases as "nation's breadbasket" and "heartland." The new farm was also close enough to the Des Moines milk market to make milking dairy cattle a reliable source of income. Although a good-sized farm for the time, it wasn't nearly big enough to make it in today's world of giant equipment, expensive genetically modified crops, and factory farms. But I'm getting ahead of the story. For now, I want to tell you about the farm as it was then. As you will see, by purest chance I was in the perfect place and time to observe what has been called the "technology treadmill" grind up one way of farming and replace it with another that can no longer be sustained. I've spent the rest of my life searching for alternatives.

The Keeney Place became part of one of the first Soil Conservation Service demonstration watersheds in 1948. A dairy farm such as ours was well-suited to conservation farming practices; in fact, it would have been considered a model of sustainable farming by today's standards. We rotated crops so the soil had a chance to rest and weeds and insects could be controlled with fewer pesticides. We usually planted corn for two years and then planted oats with alfalfa. The oats protected the alfalfa while it was getting established. Alfalfa added nitrogen back to the soil and was a key part of a dairy cow's diet. After two or three years of alfalfa, it was back to corn which, in turn, used the nitrogen the alfalfa plants had added to the soil. Most dairies farmed this way until the dairies became so large they couldn't supply all of the feed the animals needed. Soybeans, not alfalfa hay, were a more profitable source of protein for cows on the larger dairy farms. Now, few alfalfa fields remain to nourish and protect the soil.

Dad was a staunch conservationist and knew full well that sloping land was best left in pasture. But the tide of industrialization was against him. Doing the best he could under the circumstances, he carefully terraced the pasture land as a way to keep erosion in check before planting it to row crops. I was the one who plowed the last large pasture. I remember how the land resisted when the northwest pasture was plowed. Mice and garter snakes fled their long-held homes, and the deep taproots of the plants that had established life there snapped and popped as the plow hit them. Grandpa Keeney talked about breaking native prairie with a horse-drawn plow; those roots were even stronger and rattlesnakes hissed at the intruder. In the end, the extra money we made from the new crops was spent largely on the expensive fertilizer and equipment required by the new way of farming. Slowly, but surely, we learned the meaning of the old adage that farmers handle a lot more money than they keep.

The fertilizer dealers were joined by those selling farm chemicals. Commercial nitrogen, phosphorus, and potassium fertilizers became commonplace. Pesticides were developed that killed the various weeds, insects, and parasites that came calling. 2,4-D, still a popular weed killer today even though it's been traced to cancer, was our first herbicide. Dad built a tank holder that mounted on our AC-B with a pump driven by the tractor's engine. A long boom distributed the liquid under pressure through nozzles. I had the spraying job, both at home and at Grandpa's farm. I would add the concentrated 2,4-D to the bulk tank and mix it by hand, often not bothering with a mixer or even a convenient stick. I can remember the herbicide smelling of fish oil and how that aroma spread across the fields as they were being sprayed. As the spray caught the wind, I often got the backdraft. In those days, I didn't worry about cancer; I just was grateful to be rid of walking through corn in search of offending weeds.

The corn borer was our biggest insect problem. Heavy infestations can almost destroy a corn crop by making it brittle, causing the stalks to break under their own weight.

DDT, now widely discredited and banned nearly worldwide, was our first line of defense. We called in a crop
spraying airplane when the corn was about waist-high.
The low-flying plane going back and forth over our field
was of great interest to me, so I eagerly took the job of
holding the marker flag. Each time the plane flew over, I
moved with the flag so the pilot could tell where to spray
on the next pass. Of course, I got a thorough dousing in the
process. We also sprayed DDT around the dairy barn and
directly onto the cows to control flies. It seemed a godsend
not to be covered by flies when we milked. By the time
DDT was banned because it was transmitted directly to
the cow's milk, the flies were becoming resistant anyway.
The same story was playing out with the highly toxic insecticides we used against the corn rootworm. They worked
for a little while, but the rootworm soon became resistant.
This pattern of developing a new chemical, watching the
target weed or insect become resistant, and then heading
back to the laboratory to develop still another chemical
persists even today.

When we first came to the farm near Runnells, I believe we only had one tractor, an Allis Chalmers model
AC-WC. Soon we got a used AC-B for small chores. After
much pleading, I was allowed to drive the little B while
Dad put hay bales on the wagon. There were strict rules—
no shifting out of first gear, no driving faster than Dad
could walk, no turning, and no stopping. Nonetheless, I
was in Seventh Heaven. Here I was, seven years old, probably sixty pounds dripping wet, and I was grown up enough

to drive! My next plea was to be allowed to drive in first gear in first grade, second gear in second grade; well, you get the picture. Of course, that logic was soon forgotten. I quickly learned the basics of driving any machine, and Dad had the benefit of an extra farmhand. Thus began, as I am sure it is with thousands of other farm boys, a love of farm machinery. The internal combustion engine becomes a friend and a fascination, just like the animals on the farm, but even more so, because machinery is more dependable and controllable. The AC-B I learned to drive on weighed a ton or so and had a 4-cylinder gasoline engine rated at 19.5 horsepower. Its $495 price tag included rubber tires, lights, and an optional electric starter. The AC-B was an immediate success, essentially replacing the horse on the small farm. Other tractor manufacturers soon made competing models, but the patented Persian Orange AC-B was a pioneer and forerunner of things to come. Without knowing it, I honed my early driving and mechanical skills on an icon. Under his supervision, Dad let me change spark plugs and the oil, and blow accumulated plant debris out of the air screen above the carburetor. Whenever I see a little B at a farm show or fair, I want to just hug it. I'm positive other farmers feel the same way about their first tractor.

The pickup truck was an extension of the farm tractor on many farms, ours included. We even used our 1942 GMC "Jimmy" to corral the cows when we moved them to summer pasture. Once the bull took a swipe at the truck. The resulting crease in the front fender remained till the day it was sold. The Jimmy was also a great learning lab.

Its small 6-cylinder engine had lots of room around it, and the tiny carburetor was easy to reach. There was nothing fancy about it, especially in comparison to the $50,000 rigs favored by today's suburban cowboys. It had a roll out windshield, side flaps for ventilation, and a little add-on heater that added no heat whatsoever. It didn't even have a defroster. The windshield wipers worked only when coasting down a hill. It had mechanical brakes, no working parking brake, smallish tires for its size, and a too-long body that gave it more bounce than anything else. Although it had solid axles, the shock absorbers were weak, the springs worn, and it was only a 3-speed. I did everything with and to it. I changed sparkplugs and oil, repaired wiring, worked on the carburetor, greased the axles and steering, and generally learned how things mechanical worked. Only in the last fifteen years have I not worked on cars. Now opening the hood reveals a good-looking metal box and a snarl of tubes. There must be an engine in there somewhere, but I sometimes wonder where.

The tractor and pickup truck spelled the end of Grandpa Keeney's way of farming. He had two beautiful teams of draft horses, ones I loved to watch from outside the stalls. Grandpa used those majestic teams to plow, harrow, plant corn, seed grain, cultivate, mow hay and then rake it, and put it in the barn. He couldn't handle and didn't try to understand technology. Grandpa and I put up loose hay with the hayfork, a contraption that lifted the hay with a series of forks on shafts and deposited it in the wagon where we both pitched the hay to the front until there was a full load. The draft horses tracked precisely over the windrows and would stop on their own at the end of the field to be turned around when we took a break from the heat and drank cold water in the shade of the trees. Grandpa told me it was bad luck to count the cars on the train as we watched them roll by at the end of the road. In mid-afternoon we'd "drive" up to the Vandalia General Store where they had vanilla ice cream in quarts. He'd get out his jackknife, split the box, and fish out spoons he had taken from the kitchen. We sat on the store steps and savored the world's best ice cream.

Only the pressure of more production, along with the insistence of Dad and Uncle Morris, moved Grandpa to take a second look at the new-fangled world of internal combustion engines. Like most farmers of the time, he accepted cars before tractors. Grandpa used a clutch only when absolutely necessary. He just ground down the gears until they no longer resisted. I have no idea how his cars survived. He also never checked the rearview mirror, preferring to back up until he hit something that resisted his path. No one dared park close to him. He finally succumbed to the pressure of the times and took his turn on the tractor. The results were not always good. I remember one day when he forgot that our tractor had to be stopped using the clutch, and it got away from him. Several of us watched in horror as he headed toward the pond, yelling "Whoa, you son of a bitch! Whoa!!" He hit the pond at full speed. The tractor died when water hit the magneto. We fished him out, mad as a hornet, but none the worse for wear, dried out the tractor, and went on with the day.

Farms of that era are often pictured as idyllic places featuring happy families, docile animals, and sunny skies. A good bit of that was true for me, but the farm was also a dangerous place to grow up. Many a farmer and veterinarian was killed or maimed by angry bulls, and disease lurked around the corner in every animal house. One could drown in the farm pond, fall from the haymow, or be suffocated in quicksand-like piles of grain. Machinery, especially the tractor, was by far the worst hazard. In those days, tractors were hand-cranked. Turning the crank of a tractor mistakenly left in gear sent the tractor off without you, heading for whatever it could find to destroy. The crank was always in front of the tractor and woe unto you if you were in the machine's way. Other farmers perished when tractors rolled over in ditches or pitched backward if loads weren't properly distributed. The tractor's power take-off, used to drive other machinery, was always hungry for some loose bit of clothing. With one or two revolutions, legs were snapped, and worse.

I was lucky and had lots of cautionary advice from Dad. Even so, a careless moment could have made me a statistic. My closest call happened when I drove the tractor to and from Grandpa's Vandalia farm. The six or so miles had many short steep hills with sharp graveled curves. It was fun and stupid to hold in the clutch while going down the hills, letting gravity take over. The tractor accelerated to a dangerous speed almost immediately. I'd leave it in a lower gear and let the engine slow me down just before I hit the curve. Once I misjudged and let the tractor go too long before letting the gears take over. I recall seeing one of the wheels lift nearly to seat height during a reckless turn.

In the early years, machinery didn't do everything. Often neighbors pooled machinery, labor, and talent during harvest time. At first we picked corn by hand. Shucking the corn ear from the stalk, whipping off the husk, and throwing it into a wagon with a sideboard was a practiced skill that brought challenges and bragging rights. I recall the steady banging rhythm as the ear hit its target and the occasional miss followed by a healthy curse. Soon we had a mechanical corn picker that removed the ear on the go. It was much faster but, like all the other new equipment tempting Dad, more expensive.

As I mentioned earlier, we grew oats as a companion crop for alfalfa. For a couple of years, we cut mature oat plants with a special machine that tied the stalks into bundles. Then they were taken to the farm yard where the oat grains were separated from the stalks and husks, a process called "threshing." Uncle Morris's huge thresher was powered by a large gasoline tractor, but others used stationary engines powered by steam from wood, coal, or oil. Harvest lunches were the stuff of legend, as were the stories told by the sweating bare-chested crew members. I participated in a couple of these "threshing bees," still a dramatic centerpiece of most farm antique shows, and usually ran water to the crew or performed some other small task. This great rural tradition was lost when Dad got an AC-60 combine; it was pulled by the tractor and combined the cutting and threshing jobs. Dad was always adding new machinery. He pretty well had to as it was cheaper than hiring people to do the work and far more reliable. But as he bought newer, bigger equipment, he needed more land to use it efficiently. More land was something he could not afford. Dad, along with hundreds of thousands of other farmers just like him, was caught on what Willard Cochrane, one of the great agricultural economists of the twentieth century, called the "technology treadmill."

One of the great advantages of the Keeney Place was that we could sell milk in the Des Moines market. Most warm nights the milking cows camped outside under four marvelous oaks that were perched on a hill in a corner of the east pasture. The herd must have picked this spot for the view. Rolling hills to the east were suffused with fog-filled valleys, and the air smelled of earth and newly-mown hay. I often paused to daydream, knowing there was some higher being out there. As sure as I was of this, I also was just as sure that this being was not the one to whom my Swedish Baptist relatives prayed and of whom they asked forgiveness for being, after all, human. I never tired of getting our twenty or so cows in for morning milking. Their sad eyes, trusting faces, and mellow moos were constant reminders that the cow is one of the world's great creatures. By the time I came calling each morning, they were in a casual hurry to be milked in trade for their morning feed and relief from swollen udders.

I remember waking up the cows one Sunday morning and hearing the deep-throated roar of a car out of control. The noise was accompanied by the sight of a new Pontiac coupe hurtling through the fence, rolling over a full turn, and ending up by one of the oaks, smoking, but mostly intact. The cows put up their tails and ran to the pond, where they formed a huddle and watched. The driver, a fellow a few years older than me but also from Runnells, staggered out, smelling of booze. By some miracle, he showed no signs of injury except to his pride. Just then, fire burst out from under the hood. We got the hood unlatched so I could stuff some soil around the burning carburetor. Meanwhile, the driver staggered around a bit, then jumped back in the car without a word and started it. His adventure was far from over. The accelerator had jammed full throttle, and the car roared into action. The driver bailed out, and off sped the car on a course that would have taken it through the terrified cows and into our pond. At the last moment the Pontiac veered up the hill toward the farmyard, where, thanks to an intervening tree, it came to a sudden halt. I recall the rear end lifting up as it crashed. Dad, who was running toward the carnage, took over from there while I went back to round up some very nervous cows.

Most mornings were far less eventful, and that was fine by me. Usually the lead cow got up first. A camel or a giraffe is about as graceful. First, she rocked forward on her front knees, and then maneuvered her back legs under her haunches before hoisting herself up on her front legs. With the lead cow up front and the other animals in much the same order every day, my dog Bootsie bounded alongside, no doubt irritating the poor cows as we made our way along the worn path down the hill and past the farm pond. The pond was a great place to swim naked with other buddies on a hot day and share improbable tales of our prowess. I loved to fish there, too, and we even had a small rowboat. In its early years the pond had a thriving stock of bass and bluegills. Later on, its delicate ecological balance was thrown out of whack when Dad let his city friends overfish the bass, and farmyard runoff took its toll on water quality. It was years before I learned the science of what was going on. The pond became a good reference point. Continuing on our way to the barn, the cows and I passed the locust tree where the bull once chased me around the trunk and up onto a limb. If my protector Bootsie hadn't distracted him, I never would have managed to shinny out of danger. Bootsie took a nasty hit for me and limped for several days. At the end of our short journey, the cows trooped into the barn and found their appointed places without error, just like my family staked out their pew in church and came back to it every Sunday. The gentle animals would then stand patiently, munching their individualized mix of feed, waiting to be milked.

One by one, I attached the four rubber cups of the milking machine to a cow's teats and watched as they pulsated in just the right way to coax anywhere from three to five gallons of milk from the animal's bulging udder. I had to be sure and clean the teats of any dirt and manure that had collected during the day or night before I attached the vacuum-powered milking machine. As part of this process, we often massaged the teats with a special balm that was also perfect for softening hands. Most dairy people have fairly soft hands as a result. The cows stood close together in the barn. At times it was hot and steamy between them, and the barn felt like a sauna despite a huge fan running at full speed. In ten or so minutes, it was time to move to the next cow. Some milk always remained in the teats and had to be stripped by hand. Leaving milk in the teats courted disaster because it could clot and cause mastitis, a dangerous inflammation of the udder that ruined the milk for consumption, sometimes even killing the cow.

. We were a grade A dairy, which meant that we met all the cleanliness standards required before milk could be processed and bottled for drinking. This was important be- cause Grade A milk was worth about twice as much as the lower grades used to make cheese and butter. Dairying did not guarantee riches, but when all went well it was a good living. But no matter how it went, the backbreaking work led many young farm kids, including me, to think of other ways to make a living. The real work of milking began once the milk was out of the cows. The milk from each cow weighed somewhere between 24 and 40 pounds, and the attached milking machine added another five pounds. I car- ried the heavy can to the milk house and poured it through a strainer into a steel ten-gallon milk can. When full, the milk can weighed almost 100 pounds and had to be lifted into a circulating water-cooling tank so it wouldn't spoil before the milk hauler came by each morning to haul it to Flynn Dairy in Des Moines. The work was compounded in the winter because the cows had to stay in the barn all the time. This could go on for a month or more in severe winters. Each bitterly cold day added carrying feed to the cows, pitching hay down from the overhead haymow, and hauling manure out of the barn to my list of daily chores. Our other animals required more care, too. Winter was a hard time to be a farmer.

Then, as now, weather played a key role in the life of every farmer. Storms came in without grand media announcements. There were no crawling ribbons across TV screens, the blaring of storm radios, or tornado sirens. Storms just happened. You knew something was afoot when suddenly all of nature seemed to stand still, thunder rumbled in the distance, and inky clouds rolled across a green sky. That's the way it was one summer day when Dad and I were out picking up alfalfa hay bales. The sky had looked threatening for quite a while, but by the time it became obvious this was no ordinary storm brewing, all we could do was to make a run for it to the house. I scrambled into the pickup, and Dad ran for the tractor. The hay wagon was left to fend for itself. I think Dad actually beat me back to the farmstead and the safety of our cellar basement. Fortunately, Mom and the girls were safe in town. The storm turned out to be a real doozy. We crouched under the stairs as the house rocked above us. All the while, my heart thumped in my throat as the wind howled and hail pelted down on the roof and sides of the house. Giant claps of thunder crashed again and again. Then water started running into the basement, a sure sign of trouble. Just as suddenly, the storm was over. As we gathered our wits and started out of the cellar, Dad began to look for possible damage. He didn't have to look far. I had parked the truck where I always did—under the huge maple tree in the yard. He was none too pleased to discover a large limb had crushed the bed, leaving the cab intact. Our Farm Bureau insurance wouldn't pay for a new truck, but they did pay for a used bed. From then on, the dirty green Jimmy sported a blue bed with a Chevrolet moniker painted on the tailgate.

Winter blizzards were equally dramatic. Since they often began at night, we didn't always see them coming. Sometimes it snowed without stopping for a day, or longer. The winds would howl with snow and sometimes sleet before temperatures dropped to an incredible nose-numbing cold that hurt your teeth when you took a breath. Clearing the driveway of hip-deep snow was a half-day job and, even in hardy Iowa, school usually was called off. Driving on snow and ice-covered roads was treacherous, and I often used the tractor to pull careless drivers out of the ditch. This was a dangerous undertaking, especially when the stranded driver was a bit tipsy. He'd spin the wheels violently until they caught on the gravel and come lurching toward the tractor. Most drivers expected we would pull them out for free, even though a commercial tow would have cost huge dollars. Dad told me to judge the situation; if it seemed I was dealing with a dumb driver who should have known better, he advised charging the person. Whenever I did, I was usually treated to cussing and worse. But they paid.

Hail, blizzards, and tornadoes were always dangerous

and costly, but no weather event can set a farm back like an extended drought. The 1950's drought, which I experienced as a teenager, is seared in my memory. For four long years, rain was scarce. Crops failed, animals died, streams, lakes, and farm ponds, including ours, dried up. We carried water from Runnells for the livestock; we put watering tanks on our hay wagons and pulled them with tractors, a treacherous three-mile trip. I recall the cattle bawling frantically when the odor of the water reached them. We had little hay or corn, not even corn stalks. One year the corn turned white in August and withered away. In desperation, we fed our cattle ground corncobs mixed with molasses. I don't think the farm's finances ever fully recovered from that period.

Life on the farm was much more than farming. Whenever I hear classical piano music, for example, my thoughts return to Sundays with my family. After dinner was cleared and the dishes washed, Dad usually napped and my sisters and I had homework on our minds. That's when Mom would retire to the piano. She learned to play at a young age, honing her skills at the Swedish Baptist church in Des Moines. She was an excellent pianist. Sunday afternoon was her "quiet" time, and she sometimes played for an hour or more. I often spent the afternoon listening from the nearby parlor. She played Chopin, church hymns, and really belted out "How Great Thou Art." My father, too, was musically blessed. He possessed a beautiful baritone singing voice and had considered studying opera. Then the Great Depression had come along, cutting short both of their aspirations. The farm became their living. Neither of them gave up on their music, however. Mom faithfully accompanied Dad's solos at churches, funerals, weddings, and the occasional public appearance. I must have tried every instrument available in hopes of catching on to one of them; alas, no matter how much my mother encouraged me, singing in the high school choir was a close as I ever got to Carnegie Hall.

Mom and Dad's efforts to expand our horizons went beyond music; they really tried to get us kids out and about in the world beyond our fence lines. I remember in 1952, we loaded up the Chevy Bel Air, an underpowered sedan about the size of today's Toyota Corolla, and headed west for a vacation. Mom and Dad stopped at the Runnells branch of the Hartford-Carlisle bank to cash the latest milk check, and off we went. The drought was in full force. The mid-July heat was unbearable, especially in the little Chevy with six people, including my maternal grandmother, and no air-conditioning. Mother packed a picnic lunch of fried chicken and potato salad for the first day, but hadn't counted on the early start, the extreme heat, and no good way to keep our food cool. As a result, we spent much of the next three days dealing with the unpleasant side effects that attend six people suffering from food poisoning. We stopped at the famous Wall Drug Store and the South Dakota Corn Palace as we approached the Badlands and on to the Black Hills and the Mount Rushmore National Memorial. Dad gave us a concert in the natural outdoor amphitheater associated with the Air Force Academy near Boulder, Colorado.

We stayed in a little cabin by a stream in the Rockies and went to sleep listening to the stream rapids and smelling the beautiful pine trees. Dad and I went hiking to see Helen Hunt's gravesite and became completely lost. Only luck got us back to the cabin late in the day. As we drove further west through mountain passes, the streams seemed to run uphill, a common visual trick. The drive to Pike's Peak overheated the little Chevy. We eventually made it to Yellowstone Park and saw our first geyser, brown bears, and begging donkeys. My sisters posed in a classic shot with Yellowstone Falls in the background. The picture was so great that Aunt Florence tinted it, and it proudly hung in my parents' house for years.

Lacking time and money, we headed back east. Dad had promised to pick up a soldier returning to Des Moines from the Air Force Academy in Colorado Springs. It was a big mistake. That meant seven people in the little Chevy. We stopped for a picnic dinner in Lincoln, Nebraska; it was a good time to change drivers. The soldier volunteered. We had been driving since Denver and everyone was bushed and growly. The soldier pulled out of the park and headed west. It was at least an hour before Mom woke up and screamed that we had been there before. We turned around, but we didn't have enough money to buy another meal or stay over-night. We had to grind it out to Des Moines on Highway 6, probably a twelve-hour drive. Everybody contributed to buy enough gas to make it. When we finally got back to the farm, it was still bone dry and crops were dying. The neighbor boy who'd been hired to keep the farm going had rolled the tractor, and the cows had been so mistreated they lost milk production for a long time. I felt so bad for Dad that I offered to run the farm for a week so he could go fly-fishing. He took me up on it the next spring after crops were planted. As time went on, Dad turned the farm over to me more often so he and Mom could get away for much-needed time alone together. I loved those times. There were always cows getting out, machinery breaking, and the like, but I could feel myself changing from a boy to a man.

The farm and my family were two of three great influences during the time I was growing up on the Keeney Place. The other was going to school in the nearby town of Runnells. When I was there, Runnells boasted a town of a few hundred people and featured two churches, a tavern or two, no entertainment of any sort, and a dozen businesses. In 1955, a walk through town would take you past B&M Lunch, Glenn Blood's Store, Brown Mercantile, Byers Grocery, the Farmers Grain Company, the Hartford-Carlisle Savings Bank, McGee Barber Shop, McKlveen Lumber Company, Newby Anhydrous Ammonia, Mott's DX Service Station, and Rhoten Sales and Service. I attended Runnells Consolidated School from kindergarten in 1942 through high school in 1955. My first school bus was a small Ford Model A truck sporting a cloth top, wooden side benches, and no heater. I still remember the day at school when we sat around the radio and listened to Franklin Delano Roosevelt talk about the serious nature of World War II. We drank delicious chocolate milk in little glass bottles with our snacks. Naps were my favorite part of the day, and Miss Molly was probably my favorite teacher. The years from first through eighth grade are mostly a blur. Each year we had different rooms and teachers, some better than others. Memories blend together. We had Maypole dances, suffered cold mornings on the bus, and ate tasty snacks and lunches. Our school district was far from wealthy. Field trips and what we now call "enrichment" activities were few and far between.

It seems like I spent much of my time in high school stuck in a study hall that held all four grades. Our class had eighteen students, about the number that finally graduated, though some came and went. Desks were assigned alphabetically, just as in grade school. I sat in front of and behind the same people for years unless some hapless student got moved to the front for bad behavior. The girl in front of me had pigtails, and I had no end of fun teasing her. Principal Town surveyed the scene from a desk on the riser in front of the room. The shelves behind him held the forty or so books that passed for our library. Our classes didn't use many textbooks, either. If you were not lucky enough to have checked out books from the travelling library bus, study hall could be excruciatingly boring. Like many teachers, Mr. Town seemed to have eyes in the back of his head. No one got by with anything. Once you got his stare, he would put a mark by your name indicating bad behavior. He kept track of attendance, gave bathroom permission slips, and pretty well controlled the flow of the study hall. He spoke little, taught math and physics, and kept charge of the large clock on the west wall of study hall. It ticked and tocked loudly, and its mechanical innards controlled the timers that rang class breaks, lunchtime, and so on, throughout the school.

In spite of the administration's best intentions, academic quality wasn't high. I remember our science class having an argument with a teacher who insisted that sound traveled faster than light. We challenged her to look out the window at a steam engine train coming through Runnells. We could see it long before we heard it. She decided to settle the argument by hitting a particularly inquisitive scholar with a hammer and was promptly fired. More "advanced" classes in math and physics were equally inadequate, watered down to the minimum standard of the class. English was my favorite subject. We had a teacher in our sophomore year who asked if anyone planned to attend college. My best friend Bob Cage and I raised our hands. The two of us immediately began to study not only the standard English curriculum, but also to write an endless litany of themes and other assignments. This wonderful woman, whose name eludes me, tested us just as they would in college. It helped foster my love of writing and made college English courses a snap.

Our sports program consisted of the basics. We had men's and women's basketball, and men's baseball and football. They were the only options, save for summer league softball. I'd have preferred baseball, but baseball season at school was the busy planting time at the farm, so I played on our six-man football team. Six-man football was a fast

game that required good conditioning. Everyone played
both offense and defense and only came off the field if
injured or in desperate need of a rest. I found it exhil-
arating to leave the locker room all suited up. I savored
the field lights, the cheers of the crowd, the cheerleaders,
and the crisp fall weather under a full moon. I was never
that good compared to others on the team, but it was a
great experience. Travel was fun also, especially when we
could look forward to a meal either before or after the
game. Our coach, Mr. Geery, was apparently colorblind
and had little sense of smell. Many of the guys smoked
and sneaked beer in the back of the bus. I never participated, not out of any moral sense, but because the consequences of getting caught were too severe. It soon didn't
matter anyway because Mr. Lewiston, a no-nonsense man
of greater perception, took over the program and all such
activities on the bus came to an abrupt halt.

Dad was on the School Board for several years. You
might think that would have been an honorable role in the
community, but some, especially the hooligans and their
parents, thought otherwise. We were subject to slurs in
high school and vandalism at home. I had my car vandal-
ized twice at football games, and there were many strange
cases of cows being let out, wagons moved, and fences
downed. Nights were sometimes spooky, particularly during
Halloween week when the law enforcement believed in
amnesty for the "harmless" pranksters. The damage they
did at our farm was hardly harmless. Valentine's Day was

another bad time. The crude valentines left in our mailbox hurt Mother deeply. Such meanness was a part of the community I could never figure out.

I didn't date much during high school. I was shy, for one thing, and there weren't many girls who interested me. On top of that, I was often confused by the conflict between the teachings of the strict Baptist church and the real world of dating and socializing. And, as if that weren't enough to dampen my budding love life, I carried the lingering odor of the barnyard. I dated some girls from the Des Moines churches from time to time, but nothing serious. I did manage to discover, however, that the girls from the more liberal First Baptist Church, where Dad was paid to sing in the choir for a couple of years, were much more interesting than the Swedish Baptist girls. My awkward efforts paid off way beyond my expectations when our family drove over to the town of Carlisle to have dinner at Wilbur and Lucile Goodhue's farm. They were good friends from several 4-H activities. Her brother Jim, a great guy whom I'd run around with some, took us kids for a drive so we could get away from the parents. Their daughter Betty and I ended up together in the back seat of his convertible. We had a good time that afternoon, but I knew she was dating someone in Carlisle. Years later, during our senior year at Iowa State College, I heard she had broken up, and I started courting her in earnest. I have had many wonderful experiences in my life, but none better, and more life-changing, than that first afternoon with the lovely young woman who would eventually be my wife.

High school graduation happened with ceremonies in the gym on a very hot May evening. Virtually no air circulated. Perched on risers and sweating profusely, we endured the awards and the valedictorian speech. When a person I think was from Drake University in Des Moines gave a long, boring charge to the class, we really began to suffer. Finally he uttered the words, "I would like to close." Out of nowhere came the voice of Dale Kunzie, a fine but quiet guy not known for his academics or social grace, booming throughout the crowd, "It's about time." The class cracked up. Superintendent Fleming wilted. About half the audience gasped, while the other half laughed, largely in sympathy. The speaker blushed and sat down. Diplomas were handed out, and it was over. It was a fitting end.

A few months later, I loaded the old pickup with the possessions Mom and I deemed essential to life at Iowa State College in Ames. No one was outside to say goodbye when I pulled away from the farm and headed north. After all, I was coming back next week. No one except Bootsie, that is. My longtime friend and protector watched me drive away with a sad look on his face. I suspect we both somehow knew that my life was about to change forever.

King Hall

CHAPTER 2

Learning to Be a Scientist

I sometimes think of farming methods as either "traditional" or "scientific." Growing up on the farm near Runnells had made me a pretty good traditional farmer, that is, one who learned to farm the way the previous generation had farmed. By the time I graduated from high school, I could hold my own milking cows, managing livestock, operating farm equipment, and making a good crop. I'm sure I would have learned even more had I stayed, but the world of scientific farming would never get any closer than the University Bulletins and meetings with county agents who showed us new, more modern ways to farm. Those ways of farming came from agricultural scientists working at agribusinesses and universities. A college education would be my ticket into the world of those shaping the future of farming.

Iowa State College was the only college I seriously considered. I knew the campus, especially the Memorial Union, with its soft ice cream machine and tasty hamburgers, from

4-H days. Iowa State was then a quiet, almost backwater place of about 7,500 students in Ames, a small city of about 15,000. It had well-regarded Schools in Agriculture and Engineering; students with a bent toward the liberal arts usually went to the University of Iowa, about two hours to the east. The College was attractive to us farm kids in another important way, too—classes were taught on a quarter system that allowed us to get our education in bits while we farmed during the busiest times.

My good fortune went well beyond finding a great place to study. I did my undergraduate and graduate studies during 1955 to 1965, a time I still consider to be one of the most significant decades in the history of American agriculture. I had a front-row seat from which I could watch some of the greatest agricultural changes in American history. Farm numbers dropped fast and the remaining farms were much larger. Tractors became far more common and were much larger and more powerful than the ones I had learned to operate. The vast technological system from World War II was being redirected toward farming. New plant varieties, insecticides, fungicides, weed killers, machinery, fertilizers, and large animal confinement systems changed agriculture so dramatically that farms like the Keeney Place had a tough time keeping up. Agriculture policy was also changing; more than ever, farm size and efficiency were the measures of success. Farmers were being hustled off of the farms because the government found it was cheaper to maintain farms than farm people.

I was homesick and filled with a deep sense of uncertainty during my first days in Ames. I somehow found my way to a registration area, got signed in, and was soon assigned to a temporary space with several dozen other scared freshmen. Orientation went fast with little chance to really learn anything. Within a day we went to the old Armory and had a battery of placement tests in a hot gym with only a lapboard for a desk. I took some comfort in finding my distant cousin from Forest City, Dave Fredrickson, but as it turned out he was just as scared and confused as I was. Together with so many others in our situation, we camped out in a large dining hall and waited for more permanent room assignments.

Iowa State College was in a growth spurt due to World War II and Korean War veterans returning on the GI bill. Temporary classroom buildings, early morning classes, evening classes, and Saturday classes were all part of college life as the administration struggled to keep up with the flood of new students. Living space also was at a premium, but, once again, good fortune came calling. I was placed in a three-person room at Friley-Dodge House that afforded entertaining views of trucks getting stuck in snowstorms on Lincoln Highway. Some of the more experienced residents hid beer in our wastebasket in exchange for a small portion of the spoils. Luckily, we freshman were never suspected of being part of such a scheme—otherwise, we could have been expelled for the contraband beer and you would be reading a very different book.

From the beginning, I knew I had only one chance to get a degree and that supporting me in college was draining my family's finances. I quickly became known as a serious student, probably too much so for my own social good. I always had been interested in farm equipment and set myself to the task of becoming an agricultural engineer. This should not surprise anyone, because changes in farm equipment were the most visible aspect of the technological revolution sweeping across the Heartland. I grew up in the time when much of agriculture in the Midwest was transitioning from horses to tractors. When I had the chance to eavesdrop on farmers gathered around the threshing bees, church and barn dances, and even at family get-togethers with Grandpa Keeney, the horse-versus-tractor argument often came up. Some could see no advantage to the tractor: the horse was cheaper, didn't need gasoline, and oats and grass hay to feed horses was easy to grow on poor land that should never have been planted to corn and soybeans. Others saw the tractor, convenient and ever-more powerful, as the defining symbol of modern agriculture.

Studies done in the 1950's show the cost of owning and caring for a team of horses to be equivalent to owning a typical tractor of that era. This certainly wasn't what we were led to believe in my college years; it was simply assumed without further justification that the tractor was far more efficient and the land diverted from horse feed could be put into cropland and make more money. Diesel engines made possible ever-larger tractors and harvesting machines and agricultural engineering professors regarded them as modern miracles of agricultural technology, not as the machines of war on the soil and the ecosystem that

I now consider them to be. No one even thought about how the heavy tractors damaged soils and made them more prone to erosion. Years later, I helped direct a study comparing soil compaction on a conventional farm to a next-door Amish farm that used only horses. The same soil types on the Amish farm had higher yields and less runoff.

Students in agricultural engineering started right off in math and chemistry, areas I soon realized were poorly covered at Runnells High. If not for my shirttail cousin Dave, I would never have made it through the first quarter. I'll never forget going up the stairs of Beardshear Hall to the Math 101 exam and wondering if it were my execution. I came out with a C and was surprised to learn I looked forward to Math 102. But Engineering Drawing was a different matter altogether. I did so poorly that the instructor suggested I drop Engineering as a major in exchange for a C in the course; otherwise, I would end up failing. I still recall the introductory seminar given by the Dean of the School of Engineering. He said not many of us would graduate, probably around one in ten. Alas, I was not to be among the chosen he was talking about.

Leaving the School of Engineering turned out to be great advice. The Dean of Agriculture, a true gentlemen named Floyd Andre, had an entirely different message, one that was very important for scared freshmen such as me to hear. He said about a third of us would go on to graduate school, and it behooved us to prepare academically. I had taken an elective in Beginning Soils and loved the course. Burns Sabey taught it. He was one of those incredibly thoughtful instructors who affected my life in ways unknown to me then. Sabey sent me over to the Soils undergraduate advisor, Bugs (Bruce) Firkins. Bugs was beloved by all for his advice, personal skills, and good jokes. For a while I was discouraged about my college career and had thought about going into Farm Operations, a short course that prepared men for farming. My advisor and other professors, however, quickly encouraged me to stay the course leading to a four-year degree.

I discovered that Firkins could work wonders with course scheduling. I wanted eighteen or more credit hours during the fall and winter quarters. Those two quarters offered me the most time to study and do lab work. Spring was planting time back home, so I needed a lighter load with Fridays open in order to help out on the farm. My farm background, especially milking cows at the break of dawn, made me a real morning person. This helped in the days when classroom space was limited. I took labs at seven in the morning, even on Saturdays. That left the days open for study and work. I recall some icy cold early morning walks to the temporary classrooms for lab courses. The "T" buildings were cold, poorly ventilated, and poorly lit. Air conditioning was unheard of then, so some fall and spring classes were almost unbearable. I loved central campus, a truly college-like space, not duplicated on many campuses.

I also took as many English writing courses as I could work in, even business letter writing, which turned out to be a useful course. I also enjoyed a course on the modern history of agriculture. The wonderful professor, who had a horrendous stutter, introduced us to the forward-thinking programs of Franklin Delano Roosevelt, including electrification of rural areas and the establishment of telephone lines as well as the stabilization of grain prices with long-term storage and land retirement. The Dust Bowl was part of our discussions and how the government responded with the Soil Conservation Service and its programs. We also discussed the key role of unions in promoting family-based agriculture. My father often wondered why, even in the mid-1950's, farmers voted Republican and were anti-union. I wonder the same today.

Plant genetics was also a fascinating topic. Science was just learning about the importance of DNA, and large strides had been made in the theories of inheritance, cross and hybrid vigor that make today's high yields and crop resilience possible. In 1937, the year I was born, corn yields averaged 45 bushels per acre and did not change for two decades. When I left for college in 1955, corn yields were still only 48 bushels per acre. Despite the introduction of high-yielding hybrid corn 20 years earlier, some farmers still planted traditional, home-grown varieties. I remember Betty's dad Wilbur proudly standing in a field of hybrid corn in July with the stalks way over his head. As I am writing this, Betty's brother Jim tells me his corn is yielding over three times as much as Wilbur harvested back then.

Once my major in soils was established, I naturally took a full load of courses in that field. Wayne Scholtes made soil formation, soil classification, and landscapes come alive in the classroom and especially on field trips to see what things really looked like. I learned how Iowa, which may look boringly the same to the untrained eye, is actually incredibly diverse both in soils and in geology. No "one size fits all" solution will work in this or any other state when it comes to managing a farm, or even an entire watershed, a lesson well-learned when sustainable agriculture became a workable concept. I took two advanced soil chemistry-fertility graduate courses and an advanced course in soil physics. Soil Physics was difficult, but Don Kirkham and his graduate assistant made it seem likeable and I got better grades than some of the graduate students.

My coursework in soils emphasized that soil was a chemical reactor, not something that lives and breathes, that responded well to the treatment it received before the plow came, but shuddered and fell short when overloaded with chemicals, tilled, beaten by rain and those wonderful machines, and given no chance year after year to take time off and heal. Nitrogen? No problem. Guess a bit at what comes from the soil organic matter and manure, then add whatever else that seemed necessary from a bag. Voila, high yields! In my Feeds and Feeding class, cows, hogs, chickens were given little life, only a diet that made them high yielding, fat and presumably happy, though no one broached the subject. Artificial insemination, today's standard for animal breeding, was just becoming accepted.

I still have my copy of our textbook *Soils and Soil Fertility* , First Edition, 1952. It was written by Louis M. Thompson, a mild-mannered academic who went on to be Associate Dean of Iowa State College and advisor for incoming freshman and sophomores in soil science. I knew him well and we reconnected years later when I came back to the Iowa State faculty. Thompson, along with most of his colleagues, was on the leading edge of chemical agriculture. This quote from the text pretty much says it all: "If we look at organic-matter maintenance as the criterion we can see its weakness, because if yields are climbing and organic matter levels are declining, we are more concerned with yields and organic-matter level becomes secondary. After all, our primary aim in soil management is to seek the highest yields that we can maintain consistent with the greatest profit." We also learned that the traditional sources of plant nutrients, manure and cover crops, were no longer necessary with scientific agriculture: "It has long been believed that legume crops are necessary in the rotation if yields are to be maintained. But with the development of our synthetic-nitrogen fertilizer industry, we are no longer dependent on legume crops to maintain organic matter or to maintain high yields." Both the extreme emphasis on profits and the substitution of chemical for naturally-occurring plant nutrients have become cornerstones of today's modern industrial agriculture system.

Another bastion of soil chemistry and fertility was Charles Black. Black, who taught both of the soil chemistry-fertility courses I took at Iowa State, had joined several European soil scientists in challenging the prevailing concept from a generation before, well-stated in this quote from the USDA publication *Soils and Men: Yearbook of Agriculture, 1938*: ". . . in general nothing is more vital to good soil management than providing for the regular and systematic return of organic matter to the soil." Instead, Black taught us that yields were limited by the least available plant nutrient in the soil. It was common to show the soil as a barrel with staves, the lowest stave being the nutrient most limiting yield. Soil quality and soil tilth were not even recognized. The Soil Science Society of America carried the same message, which wasn't surprising because the Society was composed of the same professors. The Society also had members of industry, and thus would be expected to be even more chemical-centric than the Universities. An example is the quote in the Foreword of the epic *Soil Nitrogen* monograph published by the American Society of Agronomy in 1965: "A major objective of modern soil science and agronomy is to put together farm management systems that will maximize the efficiency of food production and reduce costs." Notice the conspicuous absence of any reference to soil and water quality, the environment, and healthy rural communities. Other soil scientists in the US and Europe had the same message. It was taught extensively and allegiance to the gospel was critical to advancement. When I took over the presidency of the Soil Science Society of America in 1987, three decades after learning these truths in undergraduate courses, these same industrial concepts were held to be true. Even today, the message of efficiency at all costs is little diluted, but soil quality is now widely recognized as a critical parameter of yield and environmental quality.

Those were days when college costs were far lower and a poor, hard-working farm boy could walk away with a degree and only one small $500 loan hanging over his head.

Summers I worked on the farm in exchange for partial support in the fall. I never made an accounting. I probably could have done better working elsewhere; I never considered that option. I also continued the jobs I started in high school of delivering milk to the Flynn Dairy booth during the state fair and watching the tent for Stoner Piano at night. During the school year, I worked small jobs at the college when I could, especially the fall and winter quarters. I recall waiting tables in the Friley Hall snack bar, which was especially popular on Sunday nights, the time when the dorm didn't serve a meal. One beautiful fall, the most popular song was Roger Williams's "Autumn Leaves." When I hear that song, I still think of the evenings serving tables. When I became an upper classman, my work related more to my future. I did lab work and got to know the graduate students better by helping out in the soil testing laboratory and by grading some essays. I also got the opportunity to type the notes of a visiting professor, a gregarious German who spoke horrible English, but who was a fascinating person. This was my first foray into the arena of scientific publication and a chance to appreciate the diversity in science. It also made me grateful that I was one of the very few boys who had taken typing in high school.

My senior year I was delighted to win election as President of the Agronomy Student club. It was a big break. I interacted at the college leadership level and was chair of our display at VEISHEA, a spring campus-wide open house. More important, I went to the annual meetings of the American Society of Agronomy as a representative of the Agronomy Department for student activities. It was a summer meeting on the Purdue campus in Indiana. In those days, the Society was small and didn't have posters or commercial exhibits, so it was held largely in campus lecture halls or hotels. In addition to the student activities, I met some graduate students and went to some technical meetings. Those experiences, along with the encouragement of my agronomy professors, put me on the path to graduate school. Dr. Pierre, our department head, was a thoughtful, gentle person with a Door County, Wisconsin, background and University of Wisconsin degrees. Because of his influence, the University of Wisconsin was always near the top of my list of universities to consider for graduate school.

The ratio of men to women at Iowa State College was probably around two or three to one because of the lack of good majors for women, save for a world-class Home Economics program. Nursing students went to the University of Iowa. This made it tough for shy farm boys to get dates. But that was okay with me; I couldn't afford romance anyway. Betty Goodhue and I sometimes met on campus and exchanged hellos. But when we saw each other in the fall of our senior year, I could sense a difference in her smile. A friend got word to me that she'd broken up with her long-time beau. It wasn't long until I worked up enough courage for a "study date." That meant going to the library and walking her to the dorm before curfew, which during the week was early in the evening. We went to a movie on our second date. Following that, we made a trip to Des Moines to see a Broadway show. Things went quickly, and I wisely realized that Betty was the reason I never had much time for romance before I met her.

Senior year was busy. Being Agronomy Club president took some time and the challenging upper undergraduate and graduate courses pretty much tied up all my remaining hours with studying. Then came the day when I opened my little mailbox at Friley Hall and found two letters offering assistantship support for a Master's program, one from the Soils Department of the University of North Carolina and the other from the Soils Department of the University of Wisconsin. Decision time had arrived for me and for Betty. Why not go to graduate school as a married couple? Betty shifted to a degree in Home Economics Education so she could become qualified for teaching, a marketable skill that could transfer beyond Iowa. We decided North Carolina was too far from home, and I wasn't enamored with the south anyway. Wisconsin it was.

Everything happened quickly in a flurry of activity. We organized a wedding before we left for Wisconsin, looked for a place to live, took finals, and graduated. Graduation was memorable, not for the ceremony, of which I remember little, but for my pain and suffering. I'd gone back to the farm to help cultivate corn and for some stupid reason left my shirt off during the day's work in the hot summer sun. Mother treated my world-class sunburn with sour cream and chives, a spread I haven't liked since. I sort of limped back to Ames and felt thoroughly miserable during graduation. In spite of the hectic preparations for our wedding, Betty and I had time to see *Ben Hur* and Hitchcock's *North*

by Northwest at the local movie theater. We listened to
Elvis Presley and Frank Sinatra, and on Sunday nights we
would watch *Bonanza* before heading back to the campus
for another week of classes. We marveled that a sleek new
jet, the Boeing 707, was flying trans-Atlantic and Pacific
routes, while Russian satellites beeped from space. Fidel
Castro was stirring up trouble in Cuba, and we were proud
of Alaska and Hawaii, the 49th and 50th states. Buddy
Holly's charter plane crashed outside of Mason City. And
in perhaps the least heralded and most important technological advance of the year, the decade, indeed the century,
Jack Kilby and Texas Instruments invented the microchip
while Robert Noyce invented the integrated circuit.

The wedding came up on the twentieth of June, just a
month after graduation, two weeks before we were due in
Madison. I spent the day before the wedding at the Keeney
Place, putting up hay and doing the odd chore. The corn
never looked better, the pastures were green, and the cows
were happy. I found my thoughts drifting off to a place
where this idyllic scene would go on forever and be a rock
to which I could always return. But even then, the signs
that farming had turned its back on the farmers that had
made it great were hard to ignore. I could see finances were
not as good; we were still bedeviled by trying to support
too many people on the farm. Expensive new tools were
replacing the drudgery of farming, but more farmland was
then needed to pay for the tools, and there was no guarantee that Dad could afford more land. To make matters even
worse, Dad's health was beginning to slip with the never-

ending work that farming demanded. I felt badly about leaving but knew it was the only choice.

My little hometown of Runnells was going downhill, too. Plans were being made to close the junior high and high school. Dad had to drive further to find a tractor repair dealer, Mom no longer traded eggs for groceries, and the local churches were struggling to stay open. I didn't need a census to tell me that there were fewer farms: the Cutforths went, then the Harmisons and the grouchy old German up the hill. The Polhamus farm was empty, the house taken over by renters who let the house and buildings deteriorate. In 1950, when I started Runnells High, Iowa had 206,000 farms; by the time I started college in 1955, there were 195,000. The next five years claimed another 12,000.

Our honeymoon was an isolated week on Birch Lake in northern Minnesota, courtesy of Uncle Hal Quint, a successful surgeon who had built a prolific and profitable practice in Illinois. The cabin, a beautiful place on the lakeshore, had a commanding view and a new outboard motorboat. I had vacationed there before with my parents and knew the place well. Needless to say, we did not seek much outside entertainment. After an idyllic week, we came back home, put the wedding presents in the attic, packed up what we needed into Dad's Chevrolet station wagon, and set off on a course toward Madison. This time, I was leaving the farm for good. Seeing things from the distance of a lifetime away, however, it's clear that I never really left. Home for me will remain forever those rolling hills of

Iowa, the farmstead, the animals, and, of course, the family. The love of the land, of the farm, never fades.

I remember how beautiful the rain and fog seemed during our first drive around the imposing State Capitol square in the center of Madison. The city was vibrant, alive, full of traditions and ethnic and cultural choices. For two Iowans, exposed to only Des Moines and Ames beyond their cloistered farm lives, it was a bit breathtaking. We were lucky to get into Eagle Heights, a new married student-housing complex west of the campus. Surrounded by pines and hardwoods, facing the wetlands (now pretty much developed) and only a quarter of a mile from lovely Lake Mendota, it was for us like a northern resort. The housing complex was tucked into the beautiful city of Shorewood, an old and completely built-out city-within-a-city, where the houses were beautiful and society was largely based on the University social and academic life. Money was scarce, so we quickly learned the blessing of Friday fish night at the oddly-named Cuba Club restaurant on University Avenue across from the mall. Fried perch, all you can eat, crisp French fries, cole slaw and a beer—a meal fit for a king and his queen, could be enjoyed for a dollar. Betty worked for a while in a lab at the University Hospitals and by mid-October had landed a Home Economics teaching position.

Meanwhile, I became acquainted with the setup and professors in Soils and once again settled into the life of a student. The department was remarkably congenial, so much so that I remember Democrats and Republicans sitting together at our annual social and watching the Nixon-Kennedy debate on television. There were political differences, to be sure, but nothing of the mean-spirited sort we see today. We graduate students were considered common property. When a faculty member needed help with field experiments, all of us could expect to be conscripted for setting up, maintaining, and harvesting plots. This was actually fun for me. Many of the colleagues I met on a shoulder-to-shoulder basis became fast friends through my professional years. And, after all, a hard day in the sun was routine when I was in high school. Why should it be so different in graduate school, especially when the professor we were working for treated us to a good lunch and a round of beers later?

I took courses from M. L. Jackson, the pioneering mineralogist, beginning and advanced soil physics classes taught by the incomparable Champ (C. B.) Tanner, and more work in physical chemistry. I also helped my major professor, Dick Corey, with some laboratory teaching. I had soil chemistry with Ozz Attoe, the gentlest person in the world except, perhaps, for Francis Hole, who taught a terrific soil formation course. Francis, a Quaker, stood his beliefs, even with the federal tax people. My professors were generally top-notch, but my research and coursework tended to be narrowly focused on the small stuff. Because of this, I seldom caught glimpse of how what I was doing might fit in with the bigger picture. This may not be a big deal for a physicist, but it matters greatly for someone in the life sciences. I, like so many before me and so many who came after me, was being trained to reduce all of nature into small pieces of a puzzle that in turn could be changed to fit in place. We could, and often did, work for years without realizing our work had little or no practical application. That same work, however, could result in many professional papers that led to fame and fortune, at least by academic standards.

Every graduate student must complete a research project and write a thesis as part of getting a Master's degree. My project involved a lab procedure to estimate the amount of limestone needed to best raise alfalfa hay. The work I did was hardly the stuff of which Nobel prizes are made, but it got me a degree and was a good example of how scientists of the time were taking a very narrow view of our agricultural system. For too many, the world didn't extend beyond the laboratory; so-called "reductionist research" was the order of the day. To show you what I mean, let me tell you a bit more about my research, what motivated it, and what I learned.

When the first white settlers came into Wisconsin, they moved to hilly land that looked much like their native lands in Switzerland, Germany, Sweden, and Poland. They formed closely-knit communities, many of which are still strong today, and made their living with dairy farming. Alfalfa is an important feed for dairy cattle, so it became the king of crops in Wisconsin. But alfalfa needs a moderate degree of acidity in a soil. Too much acid or too much base and the crop does poorly; keep it around neutral, and alfalfa is happy. But when you farm soils, they slowly become more acid and less suited to growing alfalfa. The way around this is to spread limestone, which neutralizes the soil acids and adds calcium back in the soil. But how much limestone should you add? That's where I came in.

The soil-testing laboratory had been using largely seat-of-the-pants recommendations for the amount of limestone a farmer might need. Ohio had just developed a good way of coming up with a more scientific estimate. Ohio and Wisconsin have similar soils and climate, so it seemed reasonable to ask if the method would work as well in Wisconsin as it did in Ohio. That, in a nutshell, was the question that drove my research. The biggest problems I faced were that limestone reacts slowly in soils and it is hard to precisely measure alfalfa yields. Doing field research trials to test the Ohio method would therefore take too much time and money. So the ideal of field testing had to be short-circuited in favor of laboratory testing. I was to mix different amounts of limestone in samples of soil taken throughout Wisconsin and compare the change in acidity to predictions from the Ohio test.

This may sound simple enough in theory, but the actual work was a different story. Soils are inherently messy. They vary a lot from place to place and you can't run most analyses on a gooey soil. First, I had to be as certain as possible that I was getting representative samples from the fields. That meant taking a lot of cores, little bits of soil obtained by pushing a coring device into the ground. The standard procedure was to mix the little samples together, spread them out to dry in warm air, mix them again, and then sieve out the leaves, roots, worms, and so forth that invariably ended up in the coring device. You then ground a smaller sample in a mortar and pestle and used that soil for chemical analysis in the research project. The actual experiment involved putting the prepared samples and limestone in mason jars, hundreds of them, and measuring acidity every three months.

Even though the samples I analyzed bore little physical relationship to the soil I started with, and the soil microbes were largely kaput, there was general agreement in our profession that this was good enough. And, as luck would have it, their assumption worked well for me. When all the data were put together, Tom Richards, a doctoral student, helped me run the statistics on a huge first-generation Univac computer that ate up a massive pile of punch cards. I remember him being astounded at how good the relationships were. My degree was secure, right on the two-year limit set by the funds available. It was reductionist research, to be sure; looking back, I can see how many important questions had been pushed aside in the rush for quick, efficient answers to complex problems. But I got lucky in that the question I answered was suited to such a narrow approach. Other researchers, such as those who reduced the effectiveness of DDT to how well it killed target insects, and thereby ignored how pesticide residues harmed birds, animals, and people as they traveled through the food chain, were far less fortunate.

During our first summer at Madison, we journeyed back to Des Moines to see our parents and stopped by

Ames along the way. As fate would have it, John (Jack) Bremner, one of the world's leading authorities on soil organic matter and especially nitrogen, was visiting the department at the same time. Dr. Pierre introduced me to him. I liked this chain-smoking man with the twinkling eyes immediately. Jack said he probably would move to Iowa after he finished a sabbatical at the University of Illinois and would be starting a laboratory from scratch. He needed graduate students and post docs to run it and was looking for someone experienced and familiar with Iowa and its agriculture. I promised to keep in touch. Even though going back to Ames for my PhD was the last thing on my mind at the time, the more I thought about it, the more sense it made. It was well-known that getting both advanced degrees from the same department, while common, was not the way to get widely noticed. The additional experience offered by splitting the graduate work was encouraged. So I went through the application process and was quickly offered a research assistantship with Jack Bremner at Iowa State, which by then was a university rather than a college.

I received my Master of Science diploma on a beautiful day in a ceremony at Camp Randall Stadium. We shared a moving van with a couple from Eagle Heights who were leaving for Ames at the same time. Once the truck was loaded and on its way, we filled our car with 31-cent gasoline and hit the road. The drive from Wisconsin to Iowa along U.S. Hwy 151, the old Military Ridge Road, has become one of my favorites. I love the long slopes, strips of corn, alfalfa, soybeans and pasture. Farmsteads have silos to hold corn chopped to feed the cattle during the winter; bales of hay dot the landscape. The highway passes bars where fish fry and sheepshead (a strange local card game) dominate Friday night, the official wedding dance is the polka, and ethnic accents abound. Catholic and Lutheran churches compete for the highest pinnacle, and woe to those who leave trash on the side of the road. Climbing out of the Mississippi River valley into Iowa, the farms soon form regular patterns; strip cropping, pastures and dairy cows disappear and are replaced by rows of corn and soybeans. Erosion patterns are obvious, and as one goes east towards Marshalltown, farming straight up steep slopes becomes common. Bars become the archetypal Midwest honky-tonk variety. Gone are the little fish fry places. The experience is as true today as it was 50 years ago.

My mind wandered as we drove. Whoever thought a PhD would be the outcome when I started six years ago as a green freshman? Was this really very smart? Jobs were fairly plentiful, farming was still an option if I had the money, and wasn't it about time to get along with life? My family was no doubt starting to wonder how much longer I would be a professional student. And unbeknownst to anyone but us, Betty was pregnant. A joy, to be sure, but it made the future even scarier. Soon enough, we settled back into Ames, found a maternity doctor, and I began the long quest that would anoint me Dr. Keeney. Meanwhile, Betty happily took on that chore of supporting us. In doing so, she found a lifelong profession, teaching at the high school level, work that was every bit as satisfying as mine.

I had a lot going for me right from the start of my program. Classes I had already taken during my Iowa State undergraduate studies and my graduate studies at Wisconsin satisfied many of the requirements for my PhD. I added more work in chemistry, biochemistry, and enzymology to fill in my program. And, as it turned out, I couldn't have asked for a better person to guide my research than Jack Bremner. He was a native Scotsman, schooled in chemistry at the University of Glasgow and the University of London where he received many honors. He took a soil organic chemistry position in 1945 at Rothamsted Experimental Station, about 50 miles from London. Rothamsted, which is still a leading research program, is the oldest agricultural research station in the world. Jack was an avid athlete and played competitive tennis and field hockey. He also loved to party, and at Iowa State he and his wife Mary were well-known for their glorious parties and beautiful private dinner gatherings.

The Agronomy Department hired Jack Bremner to develop a strong program in soil biochemistry, a new area in agronomy. Soils were becoming recognized as more than fine particles with organic matter. Complex chemical and biochemical reactions were affecting crop yields and the Iowa farm leaders wanted to know more, especially about nitrogen. The department was one of the few in the country at that time to have a mass spectrometer. I will not try

to explain this instrument but you can think of it as sling-
ing molecules around in a centrifuge, thereby forcing the
heaviest ones toward the outside. The mass spectrometer
was originally invented to study the isolation of various
isotopes of uranium. Iowa State housed a federal research
lab during WWII that was instrumental in developing the
atomic bomb. Our hand-me-down from the war-time laboratory gave us an enormous advantage in learning more
about how nitrogen behaved in complex chemical reactions. So I was cast into a new arena. How could I have
predicted this in advance? Everything we did was publish-
able. I worked with Jack and some of his other students
on methods of analysis, changes in nitrogen forms, and
on how to make nitrogen more efficient for crops. It was
heady stuff for a farm boy.

One of our main partners in this quest was the Tennessee
Valley Authority. The TVA was established in the 1930's
by President Franklin Delano Roosevelt to bring hydro-
powered electricity to the poverty-stricken areas of
Tennessee, Alabama, Georgia and Mississippi. Their dams
were incredibly successful, but strong economies needed
more than electricity. So the TVA set out to revitalize farm
economies by mass producing some of the new chemical
fertilizers coming on the scene. When WWII intervened,
however, much of the TVA program was diverted to the
production of ammonium nitrate, a powerful explosive still
in use today. One of the largest facilities for capturing nitrogen from the air and converting it to ammonium nitrate
was under construction at Muscle Shoals, Alabama, when

the war ended. What do you do with an unneeded ammonium nitrate factory? Convert it to a fertilizer plant with the added component of a large research and outreach establishment devoted to advancing knowledge of nitrogen fertilizers and telling farmers how to use them.

Jack Bremner's laboratory and staff, including yours truly, became key components of the TVA nitrogen fertilizer project. Jack and TVA's top researcher, Roland Hauck, published many leading reviews that are classics today. Jack and I went to the quarterly research meetings at Muscle Shoals where I had the chance to mingle with world experts on an informal basis. Not bad for a graduate student, wouldn't you say? It was learning at its highest level, and I wouldn't have traded the experience for anything. There were, however, times when I wished that Muscle Shoals was a little easier to get to. One time we flew in on a two-engine prop plane, a Martin 404, during a severe storm. Lightning cracked and luggage spilled out from the webbed cargo hold in front of the cabin. We made it, or I would not be here, but it is the moment in my life when I was *really* glad to land. Another time, during a winter meeting, it snowed an inch or so. We watched from the deck of our motel as cars slid crazily trying to navigate the unfamiliarly slick streets. We northerners enjoyed the show.

Little Marcia arrived in 1962, and Susan followed her in 1966. The grandparents lived close enough that they could share our joy. Our housing situation improved, too, as we tried our best to accommodate the rambunctious newcomers. A small rental house just north of Main Street had a screened-in porch and provided many occasions for Marcia to play with the neighborhood children. But my success with graduate studies and happiness with my new family was in stark contrast to the decline on the home farm. Finally, on a cold and blustery March day in 1965, I witnessed first-hand the sale of the Keeney Place. Like so many farms before ours, its final rites consisted of the rhythmic voice of the auctioneer, subtle nods when a neighbor bought a coveted rake, and the smell of hamburger fried with onions for the noon break. The sky looked threatening when Betty, Marcia, and I arrived at the farm the day before the auction. A bitterly cold spring blizzard set in early the next morning, making it hard for cars to get around on the roads. I urged Dad to call a halt to the proceedings that morning, but he decided to go ahead. Few people dared brave the weather. I watched with an aching heart as the equipment, animals, and, it seemed, my memories, as well, were sold at cut-rate prices. Going once. Going twice. Sold.

As I stood there in the cold with my family watching our farm disappear before my eyes, I vowed to do what I could to help the remaining farmers. But looking back, I see that I didn't yet know how to do that. I had wholeheartedly embraced the new world of agricultural science that unfolded before me during my college years. From excelling in my first beginning soils course to becoming president of the Soil Science Society of America, and holder of most of its prestigious awards, I was a soils guy, one hooked on scientific agriculture, better living through chemistry, and, of course, machinery to apply that chemistry. I wish I had paid more attention to one of the books I hung onto. Every page, all 1,223 of them, in *Soils and Men: Yearbook of Agriculture, 1938* is full of wisdom that was soon to be plowed down by the scientists I had joined. This quote, above all others, shows what has been lost in the fog of science: "The Nation, then, looks to the future. But here there may be a conflict between the national interest and the individual interest. In order to make a livelihood now, many people find it necessary to do things to the soil that are not in their long-term interests or the interests of posterity."

Years later, I would learn the significance of those words the hard way.

Bascom Hall

CHAPTER 3

On Wisconsin

So there I was in 1965, holding a freshly-minted PhD. A new world lay before me. I was thrilled with science, especially as it applied to the world of soils. I seemed to be good at it and loved writing scientific papers. I was convinced that my career choice could make a difference in other people's lives; food scarcity issues were beginning to surface and many were concerned about the sad state of the environment. There and then I set myself on a professional road that allowed me to travel quickly, but in what turned out to be the wrong direction. The more I succeeded, the farther I got from the Keeney Place, and the more my enthusiasm turned to disappointment. That, in a nutshell, is the story of my twenty-two years as a Professor of Soil Science at the University of Wisconsin.

Like many over-trained PhD's who can only fill certain slots, I didn't have a job awaiting me on graduation day. I settled into a holding pattern as a research associate, or, more humbly, a postdoctoral student. It's a job not many like. The pay is low, but you remain in a good place to find the right job. Some find that job quickly, some after a while, and some just move on to a new career. My break came in about a year when the University of Wisconsin came calling, beckoning me back to the lovely hills, Lake Mendota, and Union Terrace. I was to be part of a not-too-well-defined push by the Department of Soil Science to establish a greater presence in soil biochemistry. I accepted in a heartbeat. It was my dream job in a community and an environment Betty and I loved. Madison was a great place to live and grow in, and, most important, it had a prestigious university and a strong department. In spite of the low pay and a job description as uncertain as the morning clouds, I felt it had all the ingredients for success. We were set. Now, I could start helping the farmers I had pledged to serve and pay back what I learned on the Keeney Place. It was 1966, and life had never seemed better.

First, we had to find a place to live. We picked Middleton, a progressive and growing community to the west of campus. Marcia started kindergarten that fall, and baby Susan began charming everyone with her smile and coos. We experienced the first of everything with the girls—walking, talking, singing, little bikes, scooters, new beds, and so on. Family bliss! Betty soon found a position in Family and Consumer Resources at Middleton High School. We embraced our new community, and it loved us back. Next, a greenhorn assistant professor (that had a lovely ring to it!) had to learn the ropes. Getting to know the department was easy. Finding my way through the maze of academia, Wisconsin style, was another story. Lowly assistant profs had a long, steep hill to climb. Publish or perish, the tried-and-true mantra of research appointments everywhere, was definitely in force at Madison. Promotions went through a tough grilling, indeed, screened by one of four divisional committees (in my case, the Physical Sciences Promotion and Tenure Committee.) I started wondering what I had gotten into, but it was too late to change course now; a lot rode on my success.

The Keeney Place, which had faded a bit from view as I took stock of the future at Wisconsin, came back in spades when I got a call at work from Mom. Dad was dead, of an apparent heart attack. It had only been a few weeks since he'd been cruelly laid off from his off-the-farm job at the Allis-Chalmers branch in West Des Moines. That had broken his heart, I am sure. First he lost the farm, and then the job. He'd been working on one of his cars, something he always loved doing, when he died. Dad's funeral was the saddest for me. His beloved Consistory Chorus, which he led for years, sang. Bill Van Zee, the Prairie City farmer whose tenor voice blended beautifully with Dad's when they sang cantatas, raised his voice in one of their favorites. In many ways, Dad's death and burial would be my final physical connection to our family farm, but it made the memory of the Keeney Place sweeter and strengthened my resolve to address the ills of agriculture that had brought about its demise. In 2002 I stood again at the same cemetery and said goodbye to Mom. My sister Patty passed away years earlier and Janis, my last living sibling, has succumbed to cancer. The nuclear family of my happy childhood is gone.

In Madison, after two years on the job, things were becoming clearer. Academic life consisted of teaching reductionist-based research, interminable meetings, and needless committees, all intermixed with farm extension service meetings and visits with fertilizer dealers. Only teaching stood between me and death by boredom. The soils department was becoming polarized in an on-going battle for direction, pitting conventional agriculture, where its traditional power and support lay, against environmental quality, which many thought was the future. Some professors were neutral and just wanted to get on with their jobs. But the change in direction was to affect all of us over the years, especially in course offerings, new hires, and facilities. The tension between agriculture and the environment was evident in all agricultural colleges and in the American Society of Agronomy, the professional society for agronomists and soil scientists.

Conventional, including industrial, agriculture is the foundation almost all Land Grant colleges of agriculture. Their political base lies in the agribusiness community of fertilizer manufacturers, wholesalers and retailers and, increasingly, very large scale farmers. Supporters for business-as-usual include international conglomerates, multinational pesticide and seed producers, and powerful statewide and national farm lobbyists, exemplified by the Farm Bureau. These groups have a vested interest in keeping the research system going as is. Research provides the industry with new products at low cost and university training yields employable graduates. Slowly and steadily, industry has also taken over farmer-level advising so that extension service personnel are encouraged to make recommendations that promote the industrial approach to agriculture. The small-scale farmer has been left out.

This fully-developed university-industry complex firmly held one corner of the vast academic boxing ring. In the other corner were national and worldwide concerns about a deteriorating environment left in the wake of cities, industry, and agriculture. In general, cities have accepted the responsibility to develop better ways of handling wastes and building cleaner communities. Of course, cities can assess taxpayers to pay the bills. Industry has responded reluctantly to laws to decrease water and air pollution. They too can recoup by passing the costs onto those who buy their products. But agriculture is caught in the middle. Agricultural pollution is mostly non-point, meaning it can't be identified and measured directly. There are many, many farmers, with some larger ventures creating a great deal of pollution. These larger operations, while technically farms, are more like industrial operations. In Wisconsin in the 1960's they were mostly potato and vegetable farms in the intensely-cropped sandy soil areas of central Wisconsin known locally as the Central Sands and some rapidly expanding dairy operations. Unlike cities and industry, neither of these groups made a large effort to clean up their acts. But environmental issues captured the nation's fancy and both political and economic forces that could at least partially offset the influence of conventional agriculture were building rapidly.

Madison was largely Democratic, liberal, and progressive. It became stridently pro-environment in the 1960's, but the rest of the state never came along. Republican Governor Lee Dreyfus in 1978 called Madison "77 square miles surrounded by reality." The split between Madison and the rest of the state carried into the University as a controversy of the roles research and education should play. In the Soils Department, the schism expressed itself most openly in the competition for department chair. A compromise was achieved that appointed first a pro-environment chair, and then a pro-agriculture chair. Both were sympathetic to the overall departmental mission; it was a matter of emphasis. The compromise failed when the pro-environment chair quit early out of frustration with the slow-moving system. The pro-ag chair, however, not only lasted the full term but eventually was promoted to become dean of the college. Walking the fine line between environmental and agriculture issues was interesting, challenging, and threatening. To be fair, over time the agricultural community accepted that there were environmental issues associated with agriculture. They just in general didn't want to do anything about it.

When I started in the department, there were only two relatively small sources of funds for research. Industry grants usually were too meager to support a full research program including a graduate student or two. Block grants, known as "Hatch Act grants," from the United States Department of Agriculture filled the cracks. Some were strictly for local issues and often widely distributed so all research faculties had at least a shot at a project. They required a minimal project description and just a short annual report. In other words, they were essentially a "gimme." Other Hatch Act funds, based on regional issues such as nitrate pollution or sewage sludge application to soils, were slightly more competitive. These not only provided graduate student support, but also were linked to similar work going on in other states. They needed regional coordination, which was planned yearly through committee meetings to share findings. These projects had some good aspects. They often were more multidisciplinary, giving a broader perspective of the research issue at hand, and I got to know colleagues elsewhere who were doing similar research. I enjoyed the regional meetings. We discussed research as well as family progress, new jokes, and gripes about common administrative problems.

"Big Science" changed this nice homey way of conducting research. The launch of Sputnik by the Soviet Union in 1957 was the turning point. Our government realized that small science wasn't going to develop the large weapons and space exploration systems we needed to compete with the Soviet Union. Big Science meant big budgets, big equipment, big staffs, and big laboratories. It also meant big payback to universities for "overhead," that is, the charge a university passes on to funding agencies to keep the lights on, so to speak. Overhead became ever more critical during the rise of Big Science. Universities built highly sophisticated facilities with no visible way to support them. They competed for high-priced faculty with no idea how they

would retain them. Big Science came to the rescue by providing both targeted research support and generous help-
ings of overhead, the financial glue that holds a research
university together. As universities pushed faculty to get the
big grants, the original mission of the "land grant" college
and university, that is, "to teach such branches of learning
as are related to agriculture and the mechanic arts, in such
manner as the legislatures of the States may respectively
prescribe, in order to promote the liberal and practical education of the industrial classes in the several pursuits and
professions in life," faded. "Mission Creep," or the expan-
sion of the Land Grant program beyond its legislated goals,
has been one of the great failures of the Land Grant system.

I worked with other faculty on some applied research involved with production agriculture early in my career, but
even that work had the environmental stamp. How could
it have been otherwise? I was a conservationist when I first
walked the pastures in the foggy early mornings to get the
cows and gazed down at the pond Dad built. That pond was
only one of the many conservation practices we used on
the farm. I had helped with constructing the terraces and
planting the multiflora rose (in retrospect, a mistake). The
landscape was my central love. I sometimes think I became
a conservationist even earlier when I visited my youthful favorite place, a small stream with a nice overhang of trees.
It was on the southern border of the farm, almost at the
Pulhamas place. There was a period when I thought little of
these moments, but during the Wisconsin years I was always
bothered by how little awareness my fellow soil scientists

seemed to have of how agriculture was taking these special places and converting them to farmland purely for profit. I was able to tie this concern to the activism on the Wisconsin campus. Many of us who became radicalized during the years of discontent went on to use that energy for something more than just making money. For several years I felt like an outsider as I questioned where agriculture was headed, decried the decline in water quality of Wisconsin lakes, and began to really push back against the protest of the agricultural lions over Rachael Carson's *Silent Spring.*

Then came the political and social tsunami known as the environmental movement. People were getting it that our environment was on the line. The many pollutants in the Cuyahoga River in Ohio caught fire and a massive oil well blew out in Santa Barbara, California. UNESCO held an ominous sounding conference on "Man and His Environment: A View Towards Survival" in 1969. Barry Commoner published *The Closing Circle* in 1971 and coined the phrase "there is no such thing as a free lunch." Closer to home, the US Army Corps of Engineers began building the Kickapoo Dam in southwestern Wisconsin, only to see its construction halted by court order when its environmental problems came to light. By the time Americans were introduced to the possibility of calamity on an unimaginable scale when the Three Mile Island nuclear power plant melted down in 1979, the environmental movement had received enough attention in my field of study that the American Society of Agronomy was publishing its own *Journal of Environmental Quality.*

That blessed my work with academic respectability, just as events were unfolding that would also grant the resources necessary to do Big Science.

To fully understand the arrival of Big Science in the world of environmental research, you must remember that as the environmental movement grew, so did activism in general. The assassinations of John Kennedy, Martin Luther King, and Bobby Kennedy affected millions, myself included. There was political turmoil and death at home and abroad because of the Vietnam War. Anti-war protests erupted across the country, especially on college campuses. Wisconsin was one of the most violent: a post-doctoral researcher died in a Sterling Hall bombing. The call for Washington to "do something" was met with waves of legislation, each accompanied by its own bureaucracy and funding on a scale suitable for Big Science. The Water Quality and Solid Waste Disposal Act came in 1965 and the Endangered Species Act in 1966. In the 1970's, environmental legislation was never on the back burner in Congress. The Environmental Protection Agency (EPA), the National Oceanic and Atmospheric Administration, and Earth Day were all established. Richard Nixon, a Republican, famously signed The Clean Air Act, and the federal Water Pollution Control Act passed despite his veto in 1972. In that same year, the Environmental Protection Agency banned DDT. In 1974, the Safe Drinking Water Act and the Wilderness Act were signed into law, followed by the Toxic Substances Control Act in 1976. Soon after, the National Energy Act became law. The EPA was established, in part, to divert environmental regulatory processes away from the USDA, which, even then, was regarded to be under the influence of agribusiness. As all of this was happening, a number of environmentally based non-governmental organizations (NGOs) were founded. Some grew quite large, others faded away. These events, and the massive funding for work such as mine, all came during my start and rise through the tenure and grant-seeking years at Wisconsin. As you can see, I was in the right place at the right time.

My first foray into the world of journalism was unintended. I was asked to give a short speech at one of the nearby University of Wisconsin experimental farms on a lovely fall day. I gave a boring speech on the nitrogen cycle that covered the rising health concerns of nitrate in drinking water. Barry Commoner's nitrate studies on the Illinois Sangamon River had stirred these concerns. I used an early generation overhead projector and spoke in a barn-like arena. I thought that was the end of it, but for some reason a newspaper I had never heard of, *The Sacramento Bee* from California, ran a story about my presentation. Nitrate in groundwater was already a big problem in California. The story was an accurate summary of my talk, but it didn't exactly please the college or its fertilizer industry supporters. The bruises were only skin deep. I recovered and actually got some unsolicited funding from a large nitrogen fertilizer manufacturer to continue my studies on nitrogen. My accidental encounter with journalism affected me in a profound way. I realized that my career path would touch on things people really cared about. I had the potential to do a lot of good science. I could affect policy and still be part of agriculture. This became, and still is, my passion, and eventually led me to Iowa State and the Leopold Center for Sustainable Agriculture.

My university position at Wisconsin involved much more, however. I was part of an exclusive fraternity. In the 1960's and 70's, it was largely a white man's club with many perks. I had retirement and excellent health coverage, sick leave, two weeks paid vacation, and a high level of respect and trust in the local, state, national, and international communities. A tenured professor keeps irregular office hours, usually has good logistical and travel support, and works on a beautiful campus stocked with happy students, good restaurants, and plenty of places to hang out. I occasionally griped about university life, but compared to the drudgery of business and government jobs, I had a very good deal. My move up the tenure ladder dovetailed with all the major environmental bills, agencies, and movements. I established research programs in seven major areas: the causes and consequences of lake eutrophication, nitrogen transformations in lake water sediments, heavy metal accumulation in soils and river sediments, nitrate in ground water, sewage sludge disposal on agricultural land, private septic tank waste disposal, nitrous oxide sources and sinks in various ecosystems, and nonpoint sources of nutrients to lakes.

Many of my programs were developed jointly with other professors, which required teamwork and good communication. Fortunately for me, the Madison campus was home to some unique interdisciplinary environmental departments and disciplines. Limnology, the study of the biological, chemical and physical features of lakes, was situated on the campus shore of Lake Mendota, said to be the most published lake in the world. The graduate school had its own water chemistry program, and the Wisconsin Sea Grant Institute focused on stewardship of the Great Lakes. Separate from these was the federally-funded Water Resources Institute. The Institute for Environmental Studies, now the Nelson Institute for Environmental Studies, provided an interdisciplinary graduate program in various environmental areas. Many faculty from different areas worked within and among these units and programs. There were also various departmental research specialties, extension service programs, and units. Even though there were times I felt that a good road map and a dictionary were needed to navigate the environmental field of the campus, I learned about related disciplines and built strong team coalitions.

In spite of my successes, I saw and thought little of the Keeney Place and had equally little opportunity for research or outreach in sustainable agriculture. At that time the sustainable concept was not on people's minds except for those who'd begun to take the Club of Rome's seminal *Limits to Growth* seriously. It never occurred to us that our use of natural resources couldn't be sustained forever. But some people *were* thinking sustainably and they would be derided by the *status quo* and a public that feared change. My own thinking shifted in the late 1960's. The course I initiated, "Soils and Environmental Pollution," helped me considerably. Two books I still own, yellowed with age but readable, also influenced me. One, *The Environmental Handbook, 1970*, edited by Garret De Bell, was prepared for the first Environmental Teach-In. The other was Barry Commoner's *Science and Survival*. When I proudly showed these books to my dean as examples of forward-thinking writing, he was taken aback. They obviously hadn't been on his reading list. Later, I was able to meet Commoner and attend some of his presentations. A powerful and intense thinker, he incurred the wrath of the agricultural community, especially the fertilizer industry, with his nitrate studies.

The family and I loved to explore Wisconsin. One of our favorite outings in spring and fall was to the Kickapoo River valley, a region tucked away in scenic southwest Wisconsin. We loved the vistas, especially around the village of Gays Mills. The apple blossoms in the spring and the soaring beauty of the hardwoods in full autumn reds and oranges were always a delight. The Algonquin meaning for the Kickapoo, "one who goes here, then there," was appropriate for the river. It weaves its way through the landscape, cutting through the limestone, sandstone, and shale for 130 meandering miles. The bluffs are mostly steep and wooded; the little bit of flood plain has been largely settled. It is heaven there, but few people know about it. That was, and is, its blessing and its curse. Put simply, it resembles an almost Appalachian area. Sparsely populated by proud people, the only arable land is on the meager hilltops and the flood plain. There is little industry, and the area is far from major population areas. It has its natural beauty, and the Kickapoo's tributaries boast first-class reproducing trout streams, but that doesn't translate into money. We always carried picnic lunches because of the dearth of good restaurants.

Those narrow canyons didn't store much water, so deluges, which occurred regularly, brought flash floods of epic proportion. The people in the valley couldn't catch a break. Along came the Army Corps of Engineers with its mandate to build things, especially dams that controlled floods. Around 1968 the Corps dusted off a 1930's plan to build a flood control structure on the river near the small town of La Farge. They proposed a 1780-acre, twelve-mile long reservoir. The Corps promised flood control for the little towns of Soldiers Grove and Gays Mills. More than that, it meant that the valley would finally have a chance to develop recreation and tourism. I followed this development in the papers with mixed emotions. I loved the river and its valley, but felt sorry for the folks who had little chance to get ahead. Madison environmentalists, especially the canoe folks, as well as the rod and gun clubs, strongly opposed the dam and they made their feelings known. In the meantime, the Corps plowed ahead and took by eminent domain 149 farms comprising 8,569 acres. Lives were shattered, homesteads of many generations lost, and a population was embittered. The project became highly controversial.

Construction of the earthen dam started in 1971, but in 1970 Congress had passed the Environmental Protection Act requiring an environmental impact statement. While not technically required of a project that had been authorized in 1962, it was prudent for the Corps to do an environmental assessment. They turned to the Institute for Environmental Studies at Madison. The Institute was funded by the state legislature and placed in a fairly independent position on the Letters and Sciences campus. It had the flexibility to use faculty from many disciplines in its research and teaching programs. Later on when I started putting together the Leopold Center in Ames, I borrowed some of the Institute's structural ideas. The Institute rapidly formed a team, developed a proposal, and began negotiations with the Corps. Four of us flew to St. Paul and met with Corps officials for a day in their regional office. We returned with a contract. The project had an economic assessment, an endangered species assessment, and a water quality assessment. I was to tackle water quality. The Corps justified the project economically on the number of visitors by assuming so many recreational visitors at so many dollars spent per day. It would be necessary to maintain good water quality—swimmable and fishable.

We were given about fifteen months to complete the report. It took all of that time and then some. I had considerable stream flow sediment and phosphorus data available for the watershed and used published data for the amount of nutrients that would flow into the proposed lake, given the land use in the area. These data were compared to published models, crude at the time but the best available, to estimate the water quality of the lake. The models showed conclusively that water quality would be low, similar to the lakes in the Madison area, which were far from pristine, and far worse than the gems in Northern Wisconsin. This critical part of the report was sent for review to experts in Canada, Europe, and Scandinavia, as well as the

United States. The Corps knew that the report, if verified, would confirm the critics and probably doom the project. Without touching a test tube or using any original data, I produced the most reviewed and discussed research I have ever done.

During the critical time before the report hit the fan, the Corps put its misguided public relations machine into full gear. They paraded us, along with reporters and locals, around the countryside in unheated school buses with completely frosted windows, fed us cold sandwiches, and belittled our work, especially the estimates of endangered species. But in the end, it was to no avail. Support for the project was withdrawn. The proponents made halfhearted attempts to get the project back on line, or, at least, build a smaller retainer structure. Over $18 million was spent by the time the project was cancelled. Property owners lost their land and the communities lost the young people who were needed to rebuild and revitalize their economies. It was lose-lose on all accounts. Half of the people blamed the Corps, the other half the environmentalists. I didn't necessarily come out as a hero. Some in the agricultural college were upset that I sided with the environmental community, and the U.S. Soil Conservation Service wasn't pleased that the report questioned dam construction. But the folks at the Institute for Environmental Studies were ecstatic at the report and my performance. I had found a second home. This was a formative experience for me. I learned the power of true interdisciplinary thinking and teamwork. No one from administration interfered with our work, we were free to do our thing, and do lots of brainstorming. Nearly two decades later I was to use that framework when forming Leopold Center issue teams.

Now, the valley is quiet and beautiful. A concrete outlet tower stands as a unique monument to the controversy, and the dam, partially completed, has reverted to natural cover. A beautiful visitor and educational center overlooks the valley, and most of the land has been turned over to a local Native American tribe to manage. Just up the road is the headquarters of Organic Valley, the largest wholesaler and distributor of organic farm products in the Midwest. The cooperative is still expanding and its products can be found in leading grocery stores nationwide. While some new homes dot the landscape, the towns remain about as they were. I'm delighted and proud that our oldest granddaughter and her husband have purchased land in the valley and are establishing the type of agriculture the land will support, called permaculture, based on grass, trees, and animals. Still, farmers were hurt and most did not return. Bitterness remains in the valley to this day. I learned that the environment really matters to people, but protecting it will often come at a cost. The Kickapoo assessment was a real turning point.

The other professional experience from those years that nudged me in a new direction was my study leave in New Zealand in 1975. I had been asked to interview for the department head of agronomy at Ohio State University. The interview didn't go well, mainly because I couldn't see how I could fit in. I turned them down, but played a little game with Wisconsin that resulted in support for a study leave. After looking around, I picked New Zealand. When I proposed we go to New Zealand, Betty was excited, but the girls were apprehensive. I didn't blame them—we were concerned, too. I had colleagues in Australia, most from connections with Jack Bremner, and seriously considered going there instead. But a dual offer from New Zealand's Department of Science and Industry with a joint appointment in the Soil Science Department at Massey University turned the trick. We didn't have computers and electronic communications, only typewriters, telephones, and snail mail, so planning our leave was not exactly easy. Nonetheless, the family arrived intact at a small airport in the small, beautiful city of Palmerston North, in what I must call the loveliest country on earth. I knew this would be my only sabbatical. Wisconsin didn't really grant sabbaticals, so mine was technically a study leave. Either way, the job of rearranging lives was something Betty and I wanted to tackle no more than once in a lifetime. But this one time was worth it, and it's an experience the four of us will never regret having. Thereafter I always encouraged my colleagues to take at least one year's leave, even though many thought it wasn't worth the effort.

New Zealand was a refreshing experience. We met honest, friendly, cheerful people. The environment was clean with beautiful scenery everywhere within a short drive. Movies were scarce and we had only three television channels, but BBC news was just fine. Stores were shuttered for the weekend from Saturday afternoon until Monday morning. The food tasted bland and our choices in restaurants and shopping were limited. As the serenity of life in New Zealand settled upon me, I began to see how I'd succumbed to the rat race of "publish or perish," chasing grants and quality students and pushing myself too hard.

Just as Iowa ran on corn and soybeans, New Zealand needed highly productive grasslands to drive their sheep and dairy. Sheep, grown for wool and meat, were everywhere. Knowledgeable Kiwis matched their animals to environmental conditions in ways lost to American industrial farmers who instead matched animals to suit the demands of factory farming. Dairy cattle were largely based on the HolsteinFriesian breed I knew from my youth, but crossbreeding with the Jersey cow produced a shorter, more rugged cow, more adapted to life on pasture. Because the weather is mild even in winter, cows were milked in the open in milking parlors, which were often portable. Many dairy farms had only a four-wheeler as their main farm tractor. Farmers adopted a system known as managed intensive rotational grazing to maximize forage productivity. This system relies on flexible electric fence, made of aluminum wire strung on lightweight aluminum stakes. Many dairy farmers in the U.S. have since adopted this approach because of the way it simultaneously cut costs for labor, feed, and equipment. The more I came to understand the rotational system, the more impressed I became, and the more I wished we had used it when I was growing up on the Keeney Place. The west pasture could have been saved from the plow. This great pasture system, based on early research at the Palmerston North Grasslands Division, relied on a specially bred ryegrass-clover mix. It fixed nitrogen at a high rate compared to pastures in the Midwestern United States. If the system was pushed too hard, that is, if too many cows were on the land cycling the nitrogen through to urine, the nitrate would accumulate in the groundwater. Without breaking open a bag of nitrogen fertilizer, nitrate pollution could occur. It was an important lesson I carried back to Wisconsin.

New Zealand was a great place to appreciate the effects of location on ecosystems and crops. Tall, specially-bred Monterey pines thrived in the sandy areas close to the volcanoes. A warm sunny valley could grow decent field corn. Ryegrass-clover pastures thrived on the hills and flatlands. The group I joined was charged with learning more about how nitrogen was fixed by the ryegrass-clover system and with figuring out just where this nitrogen was going. We started a group that met once or twice a week at tea time in the dining area of the Grasslands complex. We'd capture a table, and from four to six of us would argue a particular point or pathway until the lunch bell rang. It was a great way to get to know each other and generate new ideas. Teatime was an art. A little horn tooted to remind everyone that tea was ready, as if folks weren't already looking at their watches. I became spoiled by the Kiwi way of doing research and by life in a little slice of heaven in a mad, mad world. The Keeney family is richer for having been there.

When I returned to Madison in 1977, I tried to pick up where I left off, but it just didn't happen. New Zealand

changed my outlook on research, but I wasn't in a position
to change direction quickly and, frankly, couldn't visualize
where I might be headed. We got back to Iowa at least
twice a year for visits to Betty's family farm and other
family. Iowa continued to have a pull on our heartstrings,
but signs of trouble were everywhere. Agriculture was
undergoing another severe change. Prices for grain had
ramped up fast during the time when the Soviet Union
had elected to use its gold reserves to buy grain, and the
grain speculators went wild. In 1973 and 1974 corn prices
nearly doubled and land speculation led to large increases
in land values. In 1981 grain prices plummeted and with
them, land prices. Over-leveraged farmers were forced into
bankruptcy, driving some to suicide and many to lose all
their land. It was a sad time, and brought more emphasis on different, more sustainable ways of farming. These
changes would come too late for the Keeney Place, however. The barn was gone, no animals were in sight, and the
outbuildings were in various stages of disrepair.

Shortly after our return from New Zealand I was
elected Soils Department chair. Thus began five frustrating
years. While I think I did a pretty good job as chair, I never
knew for sure, and I was troubled by self-doubt all the time
I was in the position. The chain of command was diffuse.
A department chair is crushed between strong, demanding,
forces: largely underpaid faculty on one side and deans and
higher-level university administrators on the other. My
new responsibilities left precious little time to maintain the
quality of teaching and the level of research funding I had

before. I soon let the research slide so I could remain active in my professional societies. My work with the American Society of Agronomy brought prestige, a stronger voice in my profession, and an opportunity to serve as president of the Society. I liked the roles and challenges provided by the society and became life-long friends with the officers and the staff. But no amount of national recognition could outweigh the day-to-day frustrations of life in my department.

I began to seriously rethink where I was going. I had more than 100 professional publications, but doubted that few people beyond the reviewers had really read most of them. Not much of my research affected agriculture directly. When we would go back to the family gatherings in Carlisle, Betty's family often tried to make polite conversation by asking me what I was doing. I found I couldn't explain my work, a good test that it wasn't relevant to folks in the Heartland. I tried a couple of alternatives by becoming active in higher University leadership, serving on advisory committees and on the tenure and promotion committee. I ended up on the prestigious University Committee, a nine-person group elected campus-wide to be the intermediary between the administration and the faculty. Then I was appointed to the search and screen committee to fill the open Chancellor position and subsequently was elected chair of the University Committee, an ultimate rung on the ladder to a big role in administration. I also became active in the Institute for Environmental Studies and soon became chair of its land resources interdisciplinary degree program. I loved this role and, at the urging of the students, began work leading to a distributed degree in sustainable agriculture, a field of study just coming on the scene. This didn't please the dean of the College of Agriculture who eventually caved in to industry pressure and shut down the new degree program. It was a good lesson for me in how politically volatile sustainable agriculture was turning out to be.

As my frustration with life in Madison grew, Iowa was beginning to face up to groundwater contamination caused by farm chemicals and fertilizers. We had dealt with this controversy in Wisconsin, particularly in potato fields, but here was a new wrinkle. Several soil and water scientists, led by George Hallberg of the State Hygienic Laboratory at the University of Iowa, had established a working task force on water contamination and found funds to monitor rural wells and the output of nitrate and pesticides from the Big Spring in northeast Iowa. The Big Spring, a large artesian spring, drained a basin that had no outside sources of nitrate and atrazine (a corn weed herbicide) other than agriculture. The nitrate and pesticide levels in Big Spring echoed the hydrology of the area and provided conclusive proof that agriculture was the culprit. To top it all off, nitrate in private drinking water was often way above health limits. The public demanded action, but their demands weren't being met. A lack of knowledge of the extent and cause of the problem was exacerbated by reluctance on the part of the agricultural community to act. The chemical industry and the Farm Bureau loudly and persistently claimed that any action would curtail fertilizer use, cut yields and profits, and subject farmers to bureaucratic regulation.

Nonetheless, groundwater contamination from farming had been recognized as an environmental issue. I was one of the early researchers on nitrate in groundwater, and co-author of a 1978 National Research Council report on the topic. The research concerned particularly the porous Central Sands of Wisconsin, where potato fields required copious quantities of nitrogen fertilizer. Much of it filtered into the groundwater. I was also on a national coordinating committee for nitrate studies funded by the Agricultural Research Service. This committee met in problem areas ranging from Hawaii to California to Maine. I wrote some major reviews and by the early 1980's was well recognized as an expert on the topic both nationally and internationally. Iowa's issues were well known to me. Because 82 percent of Iowa's people depended on groundwater and the municipal water supplies of Des Moines were threatened, groundwater contamination suddenly gained prominence. The press, especially the *Des Moines Register*, carried many in-depth articles. The *Register*, widely read throughout the state, had considerable influence. Iowa had to act.

Most states approached groundwater quality legislation by establishing standards, regulations, and various methods to monitor groundwater. This was the approach Wisconsin took, and I was on a couple of state government committees that helped write the legislation. But we knew deep down that this single approach, focusing on standards, wouldn't be adequate. Many others and I have long regarded environmental standards as a license to pollute up to a given level. Iowa, on the other hand, approached the issue more comprehensively. In 1957, the Iowa Water Law was enacted to control water quality. It was a basis for the 1985 State Water Plan, the supporting document of the Iowa Groundwater Protection Strategy. The strategy importantly stressed an "ethic of prevention." This philosophy was the basis of the comprehensive Iowa Groundwater Protection Act, the legislation that brought the Leopold Center into existence and brought me closer to the Keeney Place. Major funding came from both fees and polluter pay assessments; much of the Leopold Center funding came from taxes on nitrogen fertilizer and pesticide registration. The legislation established a center at the University of Iowa to study the health effects of groundwater contaminants, and another at the University of Northern Iowa to assist in management of solid wastes. The DNR and Department of Agriculture and Land Stewardship received funds for groundwater monitoring, the closing of drainage wells, and education. The Iowa Department of Public Health would publish annual reports on water quality health issues. The approach to spread the funding around helped garner widespread support for the legislation.

Iowa State's center was the key to the legislation. Paul Johnson, a democratic legislator who lived near the Big Spring, arranged for copies of *Sand County Almanac* to be given to each member of the assembly and led the effort to name the Leopold Center for Sustainable Agriculture. Even though Aldo Leopold was far better known in Wisconsin than in Iowa, he was born and spent his first eighteen years in Burlington, Iowa. I didn't know him personally (he died when I was nine years old), but got to know and love his daughter, Nina Leopold Bradley. Nina gave lectures in Madison on the Leopold story, the Leopold reserve, the famous shack, and on restoring abused landscapes. The Leopold reserve was close to the Wisconsin River near Baraboo. It was a perfect site for measurement of nitrous oxide output from an impoverished soil, and we got permission to do some non-destructive monitoring near the Leopold Shack. I kept up with the legislation's progress through my friends at Iowa State and the important letters and opinion pieces on the *Des Moines Register*. In late winter 1987, an announcement arrived in my mailbox that was to change my life, and those around me:

Notice of Vacancy
Professor-in-Charge, Leopold Center
for Sustainable Agriculture
Position Available January 1, 1988

I applied. An insider later said that the position was written with me in mind. I qualified on all counts and had the Iowa background, to boot. I made the short list of five for interviews for the Leopold Center director. This wasn't surprising. The next month I was invited for a two-hour interview at the Des Moines airport, the typical routine when a search committee wants to confirm its choice. This was perfunctory; I flew in, talked to them and headed back to await the final offer.

Leaving a community, a position and a way of life after twenty-two years gave rise to many melancholy moments.

I recall climbing to the top of Observatory Hill. Nearby was King Hall and the Department of Soils, where the students and non-students rioted during the unsettled times of the Vietnam-Cambodian war. My office and laboratories, which could barely be seen, were boarded up after Molotov cocktails hit them during that fateful night. Down the hill to the east was Sterling Hall, destroyed by a truck bomb later that year. Bascom Hall was on the second hill, where I held forth during many committee meetings. The stately lawn east of Abraham Lincoln's statue had witnessed many student-led pranks while we were here; the one I remembered the most involved hundreds of plastic pink flamingos. Then there were the crosses of soldiers lost in Vietnam. Nearby was Science Hall, one of the oldest on campus, where the Institute for Environmental Studies had offices. Further east was Memorial Union, the first to serve beer, with its iconic terrace on Lake Mendota and the dark Rathskeller where I consumed many beers and burgers. Back to the west was Middleton, our city, where the girls grew up; now they were adults and on the way to great lives of their own. Closer in was Picnic Point, famous for its scenic walks on the peninsula stretching into the lake. Across the lake was Mendota State Hospital, where we arrived an hour late for a 10K road race one Saturday morning. Further around I viewed the Governor's mansion and, straight east, the dome of the State Capital. Wednesday night band concerts and Saturday morning farmer's markets kept us entertained during the warm summers. So much we would miss.

I have always loved the poetry of Robert Frost, especially these lines:

"I shall be telling this with a sigh
Somewhere ages and ages hence:
Two roads diverged in a wood, and I—
I took the one less traveled by,
And that has made all the difference."

It was time to pack the boxes and start anew on a less traveled path that would restore my hope and get me back to the Keeney Place.

Curtiss Hall

CHAPTER 4

The Leopold Center

*T*he *Leopold Center for Sustainable Agriculture*. What a powerful name. The Center's namesake, Aldo Leopold, a Burlington, Iowa, native, remains the icon of modern conservation thinking. An individual blessed with gifted thought and a golden pen, Leopold was a dedicated taskmaster and someone who had no qualms swimming against the current. Leopold's *Sand County Almanac* and Rachael Carson's *Silent Spring* have become the two most influential environmental books of the modern era. In the final essay of his *Almanac*, titled "A Land Ethic," Leopold shares this wisdom: "A thing is right when it tends to preserve the integrity, stability and beauty of the biotic community. It is wrong when it tends otherwise. Conservation is a state of harmony between men and the land." I inherited the responsibility to move the land ethic to the modern Iowa landscape, one of the most altered in the world. I soon found that, even in Iowa, few people outside of the natural conservation circles knew the name of their native son, and even fewer appreciated the power of the land ethic.

The word "sustainable" was introduced to the language in the late 1960s. Used as an adjective, something is sustainable when it is "able to be maintained at a certain rate or level." Often the noun "sustainability" means the "endurance of systems and processes." As the words "sustainable" and "sustainability" gained popularity in the 1990s, they were often misused and misunderstood. Many times they referred to just about anything from substituting greener vehicles for those powered by gasoline to somehow improving erosion protection in fields planted with continuous corn. Moreover, the words often were used thoughtlessly as an advertising gimmick. But when the Leopold Center was young, the word "sustainable" had power, especially when coupled with "agriculture."

The law creating the Leopold Center defines a sustainable agriculture as one that "maintains economic and social viability while preserving the high productivity and quality of Iowa's land." Perhaps Wendell Berry said it best years ago when he wrote, "A sustainable agriculture does not deplete soils or people." The first use of the term sustainable agriculture is attributed to the late agricultural economist, ecologist, and educator at the University of South Wales, Gordon McClymont. McClymont felt his undergraduate and graduate training at the Australian Universities was too narrow and hadn't prepared him for his work in veterinary science. He started the Department of Rural Science at the University of New England (Australia), which offered interdisciplinary approaches to agriculture. I've often reflected on why Australia was the starting place for such an important concept, and also on how McClymont's academic life story is so similar to my own.

In spite of a clear definition in the Leopold Center legislation, the term "sustainable agriculture" carried considerable baggage. Early on, sustainable agriculture became closely associated with organic agriculture. Sir Albert Howard and Gabrielle Howard built upon Rudolf Steiner's promotion of biodynamic agriculture and likely coined the term organic agriculture in the 1930's. The term was a reference to plant nutrients such as nitrogen, phosphorus, and potassium that were becoming available in processed inorganic fertilizers. Then, in 1947, J. I. Rodale and his son Robert (Bob) Rodale founded the Rodale Institute on a working farm near Emmaus, Pennsylvania. The program was widely regarded as the leading source of credible information on organic agriculture. I got to know Bob well and was shocked when he was killed in a traffic accident in Moscow, Russia. The accident that claimed such a lovely visionary occurred just days after a People to People tour of China, specially arranged by Bob in honor of his father. I remember all of us giving him a big send off when he departed by train to the airport in Hong Kong on the first leg of his fateful journey to Moscow.

Many vocal organic farming advocates believed their way was the only way to farm. This, of course, raised the hackles of not only the conventional agricultural folks, but also of the university and federal research communities.

Center critics often equated sustainable agriculture with organic agriculture, openly labeling the Leopold Center as an organic farming center. Most academicians refused to even look at the science of organic agriculture. Feelings often ran high and a wide communication gap divided the two camps. Even today, the catch phrase "organic agriculture can't feed the world" is the end argument for dismissing organic agriculture. My views were more centrist, but I nonetheless became associated with and was regarded as an advocate for organic agriculture. I was wedged between the proverbial rock and a hard place.

At first, many of my academic and industry colleagues were convinced I was crazy to take on the role of director of the Center. Part of my goal became to convince them they were wrong. Others had faith that I was up to the challenge and that a long-lasting organization could be established with the Iowa legislation. Most important, this latter group included the Leopold Center Board. I knew I had to move decisively, particularly during the first year. We needed a resilient, yet visionary Center that could embrace all aspects of Iowa agriculture, not just niche farmers and their markets. To be sure, we needed to develop more narrowly-focused sustainable practices that farmers could follow. A good example was the work of Alfred Blackmer, a charter member of the Advisory Board. Agricultural scientists had developed a profile nitrate test for Vermont soil conditions that closely predicted the nitrogen fertilizer needs of corn, but was not widely used because it required too much time during critical spring planting periods. Blackmer modified the test so that it better fit the Iowa farmer's needs, and the test became an excellent example of a sustainable agricultural practice. But our environmental vision had to be based on people and their farms, not on a narrow concept of practices alone—there would be no handy list of "sustainable agriculture best management practices."

Iowa was quickly becoming a land with few independent farmers. The large landowners and renters were powerful enough to do in the Center; I had to work with them, if for no other reason than I could not possibly work against them. One of my goals was therefore to build bridges with conventional agriculture, whoever or whatever that was. I continued my work with the American Society of Agronomy. As president of the organization, I had the ear of many agronomists, fertilizer manufacturers, and pesticide industry. I was instrumental in setting up and getting support for the Certified Crop Advisors Program, designed specifically for industry and extension service crop advisors. I met with the local, state, and national Farm Bureau, the Iowa and National Farmers' Union, testified before Congress and the State Assembly, co-chaired a task force with the Iowa Fertilizer Association leadership on water quality, and served as the second president of the Iowa Environmental Council. I talked everywhere I was asked and participated in countless committee meetings over the years.

I can never forget my initial engagement as Center director. Some well-meaning college organizers arranged for me to speak at the 1988 National Farm Show in Amana,

Iowa. The talk occurred before I was actually on salary, so I traveled from Madison as an invited guest. I had little idea of how to approach this, my first talk about the Leopold Center. In preparation, I wandered the grounds of the Farm Show and marveled how the machines were bigger than ever and the chemical products ever more varied. It occurred to me that this wasn't a place sustainable agriculture advocates would visit. As if this weren't bad enough, a searing drought had reached its peak in mid-August, and I was to speak in a tent with folding chairs and shade, but no ventilation. I talked twice that day. I had virtually no audience, save for one older farmer who came in to get some shade. He promptly fell asleep and only woke up when I stopped talking.

When that ordeal at last was over, the extension service kindly put Betty and me up at a farm bed and breakfast. The house was new and delightful. The owners had a 1500-acre grain and hog farm, and during the farm crisis the wife had opened the house to paying guests as a way to help make ends meet. We had a great time visiting and saw them often at future meetings. Their bed and breakfast still operates, and the farm is part of an official Visit Iowa Farms listing. It's been sustainable for twenty-six years, weathering droughts, floods, and crop disasters. This first farm visit in Iowa as Leopold Center director made me realize I had to conceptualize what a sustainable farm was all about. In this farm's case, good farming and sound crisis management have brought them through. They had no thought of organic farming and still don't, but they also have stayed away from intensive hog farming and have maintained crop diversity. I knew there were thousands of these farms in the Midwest; how could the Leopold Center help them?

These folks were the first of many wonderful sustainable farmers I met during my time at the Leopold Center. One of those farmers made an especially enduring impression on me. In my opinion, Dick Thompson, now deceased, and his wife, Sharon, should be considered Iowa's legacy sustainable agriculture farmers. They left conventional farming in 1968 and set out to find and demonstrate better ways to farm. Everything they discovered and recommended to other farmers was backed up by research conducted on their own farm. Dick and Sharon spoke widely and made many advocates. I became an ardent one after hearing them in Madison in 1986. The previous year, in 1985, Dick, Sharon, and Larry Kallem had founded the non-profit Practical Farmers of Iowa (PFI) with the express purpose of conducting collective on-farm research and demonstrations throughout Iowa. The Thompson's farm in Boone, near Ames, served as the showcase and as the organization's headquarters. I grew to know them well, spoke at many PFI meetings, and attended their ever-popular field days. I have no doubt that PFI was one reason the Leopold Center was so well accepted by many Iowa farmers.

The searing drought continued into the fall of 1988. A photographer and I went on a fly-over of central Iowa to see how the landscape had fared. It was an eye-opener as the landscape looked very different from 1,000 feet. What struck us was the brownness of the fields; the only green spots were occasional small parks, cemeteries, and the sustainable Thompson farm. Their cattle pasture, the corn, and the soybeans were withstanding the drought better than any other farms we could see. In 1993, during the height of the mid-July floods, I repeated the flight. Again, that same farm stood out. No flooded fields were to be seen on the Thompson property. It was obvious that Dick and Sharon treated the soil differently from their neighbors. They used a tillage method called "ridge till" and were careful to avoid crossing their fields with heavy equipment too often. They didn't use herbicides and kept the soil loose by cultivating. Organic matter was returned to the soil with sewage sludge from the Boone waste treatment plant and manure from their hog operation. Judicious crop rotations enhanced organic matter in the soil, too. The sludge added phosphorus and the rotations and manures added nitrogen; only potassium was needed from fertilizer. Their crops were excellent.

What we couldn't see from the air was the animal husbandry on the farm. Hogs raised in typical Iowa commercial intensive swine operations lead tough lives. They are jammed into enclosed buildings housing 1,000 or more animals in conditions that prohibit much movement. By contrast, Thompson let the hogs on his farm roam outside in lots with deep cornstalk bedding. Dick, who earned his degree in animal science, was convinced that corn grown on mega-farms was low in micronutrients. He insisted his hogs eat only corn from his nutrient-rich soils. Thompson's pigs also were exposed to a mix of beneficial bacteria through the bedding and didn't need the prophylactic drugs required by confinement-grown pigs. The bedding became compost, a perfect fertilizer for the corn land. His beef was grass fed and even then brought premium prices from those who were concerned about the nutritional quality of mass-produced beef. Later, we were to use his swine rearing concepts in our work on hoop houses.

The Thompson farm came as close to a sustainable farm as I could imagine, especially in central Iowa in 1988. Over time, PFI advocates improved on and modified Dick's philosophy. They never claimed that one size fit all, which from my point of view is a major flaw of industrial agriculture. Someday I hope to see a "Sustainable Agriculture Hall of Fame." Dick and Sharon will be the ones we idolize. Today, PFI farmers, all of whom have benefited from the Thompsons' vision of a more sustainable agriculture, cover the range of farming from one to 1,000 acres or more. I like to think the Keeney Place would have been an active PFI member.

Starting a Center or some such organization at an academic institution is usually simple enough, at least by University standards. Faculty is already in place, startup space is available, and infrastructure is present. None of this was true for the Leopold Center. Agronomy graciously offered short-term space, and a reluctant and slow-moving administration and civil service were induced to put the programs, computers, and people in place. Eventually things were up and running. The first year was so busy and at times so distracting that it was hard to do things in any "right order." It was by far the most stressful time of my professional life. Self-doubts crept in, affecting my energy level. Enough of this, I said to myself. The stress never really went away, but I learned to deal with it by taking long, relaxing runs and by going on trips away from Ames. When things really weren't going well, I had long conservations with Paul Johnson. Times talking to Paul, asking him what he would do, how he saw the issues and potential land mines facing the Center, were literal lifesavers. Paul helped me see how the Center was set up to be a source of tension with the traditional college of agriculture, both in Iowa and nationally. We were introducing a new culture. Or, was it an old culture revisited? We were to engage with academics and get them working with conservationists and farmers. We were to find ways to advance farmer profits while protecting the environment by using fewer chemicals and fossil fuels. Recruiting faculty to work on projects that didn't necessarily reflect department goals was never easy. Communication across departments and discipline lines was often difficult. One goal of the Center was to break down these barriers.

I often pondered what message the Center should project. In the end I defined it as one of hope. The Center was meant to give hope to the farmers struggling to make it; hope to Iowans that they could take their environment back and bring up their families in the Iowa of their parents and grandparents; hope to the parents who wanted to buy food they could trust; and, underlying it all, hope that Iowa State University would pay attention to the needs of the farmer and the landscape, rather than be run by its corporate sponsors. When I wanted to get off the hook, I repeated that we were on an uncharted journey, one with no timeline and no way to define when, if ever, the journey would be done. But it had to be started. It would be my job to start the journey and to build an institution with resilience, an institution with the flexibility to change with the prevailing technology, economic conditions, consumer preferences, and with the talents of its evolving staff. If I were fortunate, I had a decade to accomplish this vision.

Both Paul Johnson and I were convinced that the Center was about culture change at Iowa State University. Not just minor corrections, mind you, but big time cultural shifts. This was certain to bring about tension, not just at the administrative level, but also with those that relied on Iowa State to feed their technology and support and verify their views. In the long run, it was about corporate profits and control of outcomes. The University had been successfully turned and was no longer a people's University. It graduated students well-suited for agribusiness. Many faculty members worked with agribusiness for the betterment of corporate bottom lines and benefitted from the process with promotions, salary increases, and consulting fees. The successful faculty members were those who moved seamlessly between industry and university positions because there was little difference in what they were doing for one group or the other. The extension service had turned to providing industry information to farmers, and, as a result, lost control of its mission and, ultimately, its need to exist at all. Controversial departments, such as agriculture economics and sociology, became hotbeds of dissension as academic freedom was challenged. Many departments disappeared entirely, not only at Iowa State but at most Land Grant universities across the country, while departments serving agribusiness thrived.

The Center wouldn't be in the business of developing technology, and it wasn't meant to be a research center. We could deliver information, but not at the scale of industry or extension. We were too small to be able to make significant progress by ourselves. We needed partners to be an agent of change. As I saw it, we had to work both inside the system and within the culture and ethos of the College of Agriculture. It would be foolhardy to think we could change everything. Instead, we set a goal for the Leopold Center to make inroads in key departments. We needed to build a cadre of faculty to work on projects that were moving agriculture toward sustainability. After that we could develop good extension programming and work with graduate students through their advisors. Culture change would follow. But the Center's Advisory Board was my first priority. Initially, oil overcharge revenues had funded the Center, and the Board had run its own show for a year before I was hired. I needed to establish myself as their leader. After all, I was the director and they were advisory. The process was gradual. I made quarterly board meetings something to look forward to with good venues, nice lunches, and one or two reports on research in progress. Personal interviews of board members also appeared in the Center's newsletters.

Another early priority was to establish a reputable competitive grants program. I was glad to have had prior experience with a smaller program at Wisconsin. We sent out a request for proposals that stated the Center's mission and goals, gave applicants some idea of the scope of the projects, funds available, grant format, and the review and notification process. In this way, we sought out suitable partners and set our direction, all the while establishing our credibility. All proposals underwent peer review. The staff did the first cut, the board helped, and the rollout of funded projects got widespread attention. Everyone was watching. The projects demonstrated that the Center was open to all qualified applicants. We wanted creative and risky ideas; a reductionist approach wasn't the way the Center would do business.

By law we couldn't fund farmers directly; a state or federal agency also had to be involved. We generated a lot of interest and good publicity, both positive signs. We were picking our partners and defining in practical terms what the Center was about. We got many creative ideas, some a bit risky. Most wouldn't have been able to find funding elsewhere. Thus, a body of innovative, dedicated scientists, educators, and farmers was established, all of whom were willing to work together to bring Iowa agriculture closer to being sustainable. The Leopold Center website lists summaries of each competitive project since its inception. The information includes the investigators, funding, and the final report. Whenever I visit the site, I'm impressed again and again by the breadth and depth of the Center's competitive grant efforts. Many of the projects applied directly to farming. Some showed ways to reduce chemical use and farming impacts; others found methods to manage crop pests and diseases; some introduced new crops; and some studied low-impact ways of tillage. Sometimes a project switched specific emphasis, but the Center continued to fund a vast array of ideas that surely would have been overlooked by conventional funding streams.

I knew that competitive grants were only one of the building blocks of the Center; there had to be a way to get interactive, interdisciplinary research underway. My experiences at the Kickapoo and in New Zealand had shown me that true interdisciplinary research had four key characteristics: (1) creative ideas and interactive investigators, but one strong leader; (2) a central focus with a lot of potential avenues of research; (3) a place and funds to meet and toss the ideas around; and (4) an open source of information with critical input from vested stakeholders. I set out to create "issue teams" that met these criteria. Issue teams weren't a new concept, but at the time were largely untried, especially in agriculture. The staff and I developed a set of guidelines for structuring these issue teams. They involved funding a leader at a level that wouldn't penalize his or her basic research program and would provide assurance of long term funding and flexibility in how the work was carried out. The team had an additional budget for hiring students, funding their own meetings, travel, supplies, and other needs they deemed necessary. Each team had to have a mix of scientists, conservationists, and farmers.

I met with College administration, key faculty in the College, and the Advisory Board. Once all parties agreed on the issue team concept, I called a meeting of people whom we felt might be interested. I was pleasantly surprised to see how many came and listened. We'd sold the faculty short—many wanted to spend part of their careers in sustainable agriculture and were honored to be considered by the Center as potential leaders. The faculty most interested met with us one-on-one. Their proposals were discussed, modifications offered, and thus began the process of setting up the teams. Each interview was held individually and all proposals were discussed with the Board. The process was lengthy, but worth it. I told the Board that it would be difficult to judge success and failure.

Each team advanced sustainable agriculture in its own way, and each developed comradeship among a group of faculty, farmers, and graduate students who understood what sustainable agriculture was about. They did this on their own and then went on to tell others. Many sustainable agriculture researchers and teachers emerged from the Center teams. All they needed were the resources and some initial direction, and then we got out of the way and let them get to work. The teams proved their worth many times over. Team members also qualified for individual competitive grants and, often, team goals were supported by significant government grants. Some of the teams were especially successful. The rotational grazing team showed that beef and dairy can be fed on Iowa pastures while improving the land. The nutrient management team worked on use of nutrients from swine houses in a safe, economical way; and the pest management team improved on integrated pest management in search of alternatives to pesticides. As teams went on to garner support from other agencies, they greatly expanded their ability to address key issues in sustainable agriculture. These big trees nurtured from little acorns have put Iowa State University on the map for divergent research and outreach and established the Leopold Center as a premier sustainable agriculture funding agency in the world.

Of the issue teams I worked on, the hoop group and the riparian waterway teams stay in my mind. In Iowa, hog confinements are the norm. They are huge, cause health issues with neighbors, and have taken away another option for the family farmer struggling to compete with large industrial operations. With the hoop houses, the Center attempted to make a positive statement on confinement alternatives, and succeeded. The hoop group, started in 1992, consisted of dedicated scientists, farmers, and extension people who explored the use of hoop houses for sows, pigs, and for finishing hogs. The hoop structures were metal circular frames covered with tarp and looked much like a Quonset hut. They can be found on Iowa farms today. I loved to see the happy hogs and their piglets nestled deep in the corn stalk bedding. On a per hog basis, the returns were about the same as from the huge intensive swine units. It was a much more humane way of raising hogs. Manure is composted and nutrient losses are less than in swine confinement systems. Hoop houses continue to be used for beef and for hay storage during off seasons. I'm convinced that this simple, though elegant, technology could have kept many Keeney Places away from the auction block.

The riparian buffer project was the first issue team. It focused on Bear Creek, a steam resulting from the drainage of farm fields that had been "tiled." Underground tile lines drain water that needs go somewhere. The "somewhere" is streams that have exceptionally high erosion along their banks. The team began by planting trees and grass along the eroded stream. Water clarity improved rapidly and continuous monitoring has shown sediment in runoff being reduced by 70 to 95 percent. Another benefit was that soil organic matter and infiltration increased and stream bank erosion declined significantly. For nearly a quarter of a century we've been learning from Bear Creek and it has become a nationally-recognized USDA demonstration site. I still take a drive by there once in a while, often taking visitors to see the site. Every time my thoughts wander back to when it was a degraded stream and we were standing on an old bridge visualizing the project.

I was convinced early on that the Leopold Center model could be replicated elsewhere; I saw myself as the "dean" of the sustainable agriculture directors. We would be mutually supportive and sponsor joint programs. (The Leopold Center could not fund any projects outside of the state.) During the time the Center was established, many other states considered establishing sustainable agriculture centers, though using the "sustainable" moniker had to be avoided at times for political purposes. The ones that did get established were in response to farmer demands and faculty desires, but since they were all creatures of the agricultural colleges, they haven't been adequately funded and are subject to the whims of university budgets and administration pressures. Only the Leopold Center, with its mainline budget from dedicated funds from taxes on inputs, has held reasonably constant. Even though I got along well with the fertilizer industry, they would never again allow such taxes to be passed. I did what I could to help persuade other states to take the Leopold Center's funding route, but to no avail. The door was sealed.

The Dean of the College of Agriculture and I took our jobs at about the same time. We both had ties to Wisconsin and had much in common. He took considerable pains to keep me in his loop the first few years and was very helpful. He introduced me to the Farm Bureau Board at a lunch meeting, which made working with them easier. We and our wives got together socially. He even helped me exchange desks so that my office could have a Leopold desk from his office. But somewhere along the line there was a marked cooling of relations; I will never know why, but I was disinvited from key committees and he became difficult to work with. The same was true of the Associate Dean. It was widely known that we weren't getting along, but because I couldn't figure out what I had done wrong other than assert our position on key sustainable agriculture issues, I just shrugged my shoulders and moved on. Differences between sustainable and industrial agriculture will always bring tension and conflict to an established college of agriculture. I saw it first-hand at Wisconsin, too, and at Iowa State with Leopold Center directors who followed me. I've often declared the biggest enemy of sustainable agriculture is the agricultural college itself.

There were several divisive issues that may have affected my relationship with the agricultural administration. Hypoxia was one of the most contentious. In the late
1970's Gulf Coast water scientists, in cooperation with the
U.S. Geological Service, began monitoring the water quality of the northern part of the Gulf of Mexico, checking
mainly for the amount of nutrients, algal growth, and the
level of dissolved oxygen. This region, stretching from the
mouth of the Mississippi River to about Galveston, Texas,
was witnessing strange behavior in crabs and fish, as well
as a high die-off of bottom-dwelling ocean life. Soon it was
discovered that the oxygen dissolved in the bottom water
in this region was very low, and a threat to life in this critical zone. The low oxygen function was termed "hypoxia"
from a medical term, indicating some oxygen, but not
enough to sustain life. Scientists soon traced the symptom
to consumption of oxygen when algae died and fell to the
ocean floor. Oxygen from the top waters did not mix with
the lower level. Sea life moved away from this area if they
could; those caught on the bottom died. Hence the area
was known in the press as the "dead zone." And the source
of the nutrients that lead to high algal growth was traced
back to the Mississippi River and upstream to Corn Belt
states, especially Iowa.

It turns out that nitrate-nitrogen is the big culprit. We
had long known that the intensively cultivated and drained
soils in Iowa were major sources of nitrate to streams and

rivers, much of which ended up in the Mississippi River, which drains 44 percent of the United States. And, of course, fertilizer was suspected as the major perpetrator. Industry and agribusiness went into a typical deny, attack, and confuse mode, and many Corn Belt Land Grant universities were hesitant to step in for fear of losing support from industry friends. The issue seemed ideal for the Leopold Center to study. The answers required science and outreach from many disciplines, landscape level studies, and real team approaches, similar to what were used in the New Zealand and Kickapoo research. But Iowa really dug in its heels. Even the Iowa Secretary of Agriculture denied that hypoxia existed. The University stayed on the fringes, and I was discouraged from Leopold Center involvement. Only Minnesota and Louisiana land grant universities seemed to be willing to form coalitions. Reducing the amount of nitrate to the rivers and streams turned out to be very difficult, given the host of economic, political, agronomic, and weather related factors involved. Nitrate levels have not declined and the area in hypoxia has not changed (after allowing for changes in rainfall and stream flow). Changes will not come until landscapes have been altered to more natural sustainable configurations. The Leopold Center carried this message in its columns and speeches, as did many other environmental groups. But the Center was stymied by outside pressures from putting together the teams needed to address the larger issue.

Hypoxia was not the only wedge separating us from agribusiness interests. I was Center Director during the time genetically modified organisms (GMOs) invaded mainstream agriculture. The chemical giant Monsanto developed the herbicide glyphosate, trade-named Roundup, in the 1970's. Roundup is a broad-spectrum herbicide, which, unfortunately, kills not only obnoxious weeds but valuable crops and just about anything green it touches. In the early 1980's, the company's genetic engineers began work on inserting a bacterial gene into plants to see if they could get tolerance to Roundup. They found a strain of bacteria that thrived in their glyphosate waste ponds and successfully transplanted the resistant gene into soybean plants. In quick order, USDA gave its stamp of approval to the GMO newcomer, and so-called "Roundup Ready" soybeans were widely accepted by U.S. farmers. The use of genetic engineering to produce a transgenic crop (i.e., a crop containing a foreign gene) created major ethical issues and, since the technology was patented, the opportunity for farmers to save seed for the next year's planting was lost. In a *Leopold Letter* column, I spoke out against GMO technology. This wasn't a popular position to take at Iowa State, where most thought it would cut farmer costs and lower pesticide use. Today, only three companies dominate the seed industry, seed prices have soared to astronomical levels, many weeds are resistant to Roundup, and pesticide use has increased markedly. Many predicted this outcome, but to no avail. Once again, industrial agriculture found an unsustainable model.

Another contentious issue was ethanol production from corn. As corn yields climbed, it was obvious to many that Iowa had to develop more markets for this surplus corn. Increasing exports was one option, but grain exports have always been iffy, and remain so today. We needed a good, reliable domestic market for corn. Feeding corn to meat and dairy animals and chickens was a mature market, and not likely to increase much. Even corn processing was a flat market. A person can eat only so much food. Then, along came ethanol. In a brilliant move, corn processors proposed that the United States develop policies to establish ethanol in gasoline as a national requirement. Proponents held out the prospect for cleaner, safer fuel, and decreased reliance on oil imports, especially from hostile countries. Their arguments found support in Washington, which required national gasoline distributors to blend ethanol in their gasoline. Iowa led the nation in corn production and soon it led the nation in processing plants converting corn to ethanol. I was never enamored with ethanol for automobile fuel. Many studies, including one of my own, had shown it took about as much energy to make ethanol as we got back. Furthermore, the increased demand for corn made conserving the land and protecting the groundwater even more difficult. But the country was sold a bill of goods and soon few dared speak against this darling fuel. I was one of the few and got chewed out by the agricultural establishment on a regular basis.

Aldo Leopold once said, "We shall never achieve harmony with the land, any more than we shall achieve absolute justice or liberty for people. In these higher aspirations the important thing is not to achieve but to strive." A decade into the Center leadership, I was beginning to understand what he meant. In spite of the Center's many successes, I

more often saw how Keeney Place agriculture continued
to slip. Farmers left the land, erosion continued unchecked,
and nutrient levels in rivers, lakes, and streams were higher
than ever. The state wasn't gaining back its natural lands.
Biodiversity was never so threatened. I was losing my effectiveness. It was time for someone else to take the reins. I an-
nounced my retirement quietly, and I picked age sixty-two
as the target. It slipped by a few months, but no problem
there. I'll never really know whether it's best to just retire
on a high point, or to give some notice and be a lame duck.
In any case, I chose the latter. I've been amazed repeatedly
at how academia takes such a long, agonizing time to name
replacements. The basketball or football coach can be fired
or quit and within a few days a replacement is named. Yet in
academia, even when a long lead-time is given for replacement or retirement, not much goes on until one cleans out
his or her desk. Then an interim is named, someone with no
power to do much of anything, while the laborious search
process begins. On January 1, 2000, I no longer was an official employee of Iowa State or anyone else, for that matter. I
retired by popping champagne at a New Year's party. Since
shortly after our wedding in June 1959, when I started at
Wisconsin, I'd been on someone's payroll. My new jobless
status felt good, yet strangely disquieting.

The Leopold Center continues on in ways that restore my hope for the future of American agriculture. It
is exploring locally sourced foods for educational institutions, including the University of Northern Iowa and Iowa
State University. Community Supported Agriculture has

also been a boost for the niche markets. The Sustainable Agriculture Graduate Program has opened up new and exciting possibilities for graduate students to receive diverse advanced degrees in the many areas of sustainable agriculture and has helped faculty develop research and teaching programs outside their disciplines. The Center's work with the Practical Farmers of Iowa is now integrated to the extent that Iowa State has many joint programs. Even extension has joint appointments today. A strong interdisciplinary research program has been established in Agricultural Ecology, as well. All these successes, and many others, continue to bring state and national attention to the Leopold Center. I am proud to be one who was involved at the start.

There is still more to do, always will be. The Center is working against huge forces that will continue to try to corporatize agriculture. The journey will continue into the uncharted future. The problems I foresaw in the 1990s remain and are magnified. The delicate balance between ethanol production and the environment has not been resolved. Genetic modification, gene patenting, the increasingly powerful seed industry, environmental issues, and human health are ignored by the Land Grant universities. Corporate dominance in the overwhelming number of large confined animal feeding operations (CAFOs) continues to increase. The industrial model will fail, not catastrophically, but slowly, one piece at a time, and sustainable agriculture will be there to help put it all together again. The Keeney Place will, I am sure, be part of that renaissance.

EPILOGUE

Our oldest granddaughter was married on a farm over Labor Day weekend in 2013. Maureen and Peter said their vows under three beautiful oak trees. Lush pastures spread in every direction. Cattle grazed in the background, pigs oinked contentedly as they enjoyed a favorite mud hole, and guinea hens and turkeys foraged nearby. The dogs stayed close to the happy couple as the guests, including four generations of family, brushed away tears of joy.

After the ceremony, everyone gathered for a feast grown entirely on the farm. A spectacular sunset over the hills and valleys gave way to an incredible starry sky. I thought back to the farm of my youth, the Keeney Place. I thought about the wonderful pastures there, and how I had led our cows on parade to be milked. I wished so much that our farm had turned out like this one where grassy pasture still nurtured the soil.

My sadness over the home farm slowly gave way to hope for the future. There are better ways to farm, ways that sustain our soil, our water, our health, and our rural communities. A new generation of farmers is redefining sustainability. There are places like the Leopold Center dedicated to providing the science and appropriate technologies those new farmers will need. And there are millions of people like you who care about what happens to our farms and our food system.

I like to think I had a small part in helping this new movement get its footing. If that is the case, my quest to return to the Keeney Place of my youth will have been fulfilled.

DENNIS KEENEY

AMES, IOWA

DENNIS R. KEENEY is emeritus professor of Agronomy and Agriculture and Biosystems Engineering at Iowa State University. He has degrees in Soil Science from Iowa State University and the University of Wisconsin-Madison, and was on the Soil Science and Water Chemistry faculties at UW-Madison from 1966–1988. He was the first Director of the Leopold Center for Sustainable Agriculture, Iowa State University. He is past president of the Soil Science Society of America, the American Society of Agronomy, and is a Fellow of the American Academy for the Advancement of Science. Since his retirement from Iowa State in 2000, he has held positions as Senior Fellow for the Institute for Agriculture and Trade Policy in Minneapolis and the Department of Soil, Water and Climate, University of Minnesota, St. Paul. He has also served on the Boards of the Thomas Jefferson Agriculture Institute and Food and Water Watch. In 2008 he was appointed Visiting Scholar, Center for a Livable Future, Johns Hopkins University. Dr. Keeney and his wife Betty travel widely and are sometimes found in Ames, sometimes in Minneapolis, and sometimes elsewhere.

PAUL W. JOHNSON served in the Iowa State Legislature from 1984–1990, was Chief of the USDA Soil Conservation Service from 1993–1997, and served as the director of the Iowa Department of Natural Resources from 1999–2000. In the Iowa Legislature, he gained recognition as a leading voice for sustainable agriculture and led the effort to pass Iowa's milestone ground water legislation, which established the Leopold Center for Sustainable Agriculture at Iowa State University.